Soft Computing

Springer
*Berlin*
*Heidelberg*
*New York*
*Barcelona*
*Hong Kong*
*London*
*Milan*
*Paris*
*Tokyo*

Andrea Tettamanzi • Marco Tomassini

# Soft Computing

## Integrating Evolutionary, Neural, and Fuzzy Systems

With 81 Figures and 5 Tables

Springer

Andrea Tettamanzi
Information Technology Department
University of Milan
Via Bramante, 65
26013 Crema (CR), Italy
E-mail: andrea.tettamanzi@unimi.it

Marco Tomassini
Computer Science Institute
University of Lausanne
1015 Lausanne, Switzerland
E-mail: marco.tomassini@iissun4.unil.ch

Cover picture by Jantje Janßen, Karlsruhe

Library of Congress Cataloging-in-Publication Data

Tettamanzi, Andrea.
Soft computing: integrating evolutionary, neural, and fuzzy systems/Andrea Tettamanzi, Marco Tomassini.
p. cm.
Includes bibliographical references and index.
ISBN 3540422048 (alk. paper)
1. Soft computing. I. Tomassini, Marco, 1949– II. Title.
QA76.9.S63 T48 2001
006.3–dc21 2001042073

ACM Subject Classification (1998): F.1, I.2, D.1.2, I.5.1

ISBN 3-540-42204-8 Springer-Verlag Berlin Heidelberg New York



Springer-Verlag Berlin Heidelberg New York
a member of BertelsmannSpringer Science-Business Media GmbH

http://www.springer.de


Printed in Germany



Typesetting: Camera-ready by the editors
Design: design + production GmbH, Heidelberg
Printed on acid-free paper SPIN 10797552 06/3142SR – 5 4 3 2 1 0

# Preface

THIS is a book about soft computing and its applications. Soft computing is a general term covering a number of methodologies and, as such, it is hard to come up with a neat definition of the meaning and the boundaries of the disciplines involved. A first distinction must be made, in our opinion, between soft computing and *computational intelligence*, since both terms are sometimes used as synonymous. To us, computational intelligence has a broader scope and encompasses all the techniques that can help in modeling and describing complex systems for inference and decision making, including soft computing. Thus, AI approaches such as knowledge representation systems through crisp logical rules, symbolic reasoning, and computational language processing among others certainly belong to the computational intelligence toolkit but do not qualify, in our view, as soft computing proper. What is soft computing, then? The common thread through all the soft computing methodologies is that, unlike conventional algorithms, they are tolerant of imprecision, uncertainty, and partial truth. Soft computing techniques do not suffer from the brittleness and inflexibility of standard algorithmic approaches. As a consequence, they offer adaptivity, i.e., they can track a changing problem environment quite well. Given that many problems of interest do not admit solutions that are fixed once and for all, the adaptive character of soft computing is a definite plus. According to this loose definition, a number of disciplines such as rough sets and probabilistic networks would qualify as being of soft computing type. But the book could not be all-inclusive and we were confronted with a choice. We made the rather arbitrary decision of covering only a subset of these, namely, fuzzy logic, evolutionary algorithms, and artificial neural networks and their various combinations. The reason is that it is easy and natural to combine these disciplines synergetically and they are most useful when used together, a fact that has been demonstrated in practice by a large number of successful applications. These methodologies can be said to be bio-inspired. Thus, "natural computing" would have been another possible title for the book, but we wanted to stick to established terminology to avoid confusion. Even so, each isolated technique would require a book of its own to be fully explained. We do present the basis of each of the approaches rather extensively in the first part of the book but, since the main theme is their integration, in the second part these same components reappear under various disguises, thus strengthening the understanding of the whole. In this sense, the book is self-contained.

At this point, it is only fair to acknowledge that there has been some argument in the literature as to what should count as soft computing and what the unified bases of the discipline are, if there are any. Such concerns have been expressed for instance by Dubois and Prade [57]. Dubois and Prade claim, with some reason, that soft computing does not constitute a scientific "discipline" in the philosophical sense, since the different techniques that are lumped together under this name do not share the same foundations and have different concerns. In principle, we agree with this argument, since the three main strands of soft computing described here do not have much in common as they get their inspiration from different paradigms. These are biological nervous systems for artificial neural networks, selection in natural populations for evolutionary algorithms, and approximate reasoning for fuzzy systems. However, common features of these techniques are not to be seen in their foundations or sources of inspiration, but rather in their common approach to problem solving. Thus, if we shift the emphasis from a rigorous epistemologic view to a more goal-oriented and pragmatic one, we can see that there are indeed many links and commonalities. If we look closer, we will notice that fuzzy logic is a formally sound way of including and dealing with imprecision in data and reasoning. Artificial neural nets also admit a degree of imprecision of a different kind and at different levels: data used to train the nets can be noisy to some extent without affecting learning too much. As well, a trained neural net should be capable of working in the face of data variability and perfect discrimination power is not required. Evolutionary algorithms, once a suitable encoding for candidate solutions has been found, work without explicit knowledge of the problem structure and tend to come up with satisfactory solutions rather than optimal ones. Therefore, although the approaches look different, there is an underlying commonality of intents which, in our view, is enough to justify treating soft computing as a problem solving discipline, and when the three main components are suitably combined this is even more evident.

Thus, soft computing as we intend it here is not just a random mixture of these ingredients, but a discipline in which each constituent contributes a distinct methodology for addressing problems in its domain, in a complementary rather than competitive way. The guiding principle is to exploit the tolerance for imprecision to achieve tractability, robustness, and low solution cost even for difficult problems. We are concerned with the integration and cooperation of the main soft computing techniques among themselves rather than with their "serial" aspects, in which one methodology is used as a pre-processor or a post-processor of another, although this simple addition can also be useful at times.

In recent times, soft computing techniques have found their way into important industrial application fields such as optimization, decision support, process control, forecasting, and identification and they constitute a sizeable part of the huge appliances market, especially under the form of neuro-fuzzy hardware.

To our knowledge, the kind of integration that is attempted in the present book has never been done before as a monograph. The book is intended both for undergraduate students in soft computing disciplines as well as for engineers, practitioners, and problem solvers in many areas of application. Some parts of it could also be useful for a senior-level course. The book is structured into two main parts and its organization reflects the basic structure of soft computing as it is intended here. In the first three chapters we describe the fundamentals of evolutionary computation, artificial neural networks, and fuzzy systems. The presentation is compact but reasonably complete and it should allow a good grasp of the fundamental notions in these three fields. Readers with a good knowledge of these subjects could skip one or more of these chapters or go directly to Chapter Four. The second part, including Chapters Four to Eight, is the main theme of the book and it consists of the study of the various ways in which the three fundamental branches of the soft computing tree interact in a cooperative way so as to enhance their respective strong points. Chapter Four deals with the evolutionary design of artificial neural networks. The subject goes back to the end of the 1980s and it is thus relatively new. Most of the material is scattered in different publications and it has probably never been presented before in a unified way apart from in review articles. Chapter Five is about the use of artificial evolution in the design of fuzzy systems, which is also a subject that has not been paid enough attention to in spite of the growing number of applications. Chapter Six, on neuro-fuzzy systems, is more classical and contains the basic ideas in the field. No discussion of soft computing would be complete without it, given the commercial importance of neuro-fuzzy applications. Chapter Seven, on fuzzy evolutionary algorithms, is in some sense the reciprocal of Chapter Five and deals with the use of fuzzy systems in artificial evolution. The idea is to use fuzzy principles to dynamically tune and monitor artificial evolutionary processes. This side of the fuzzy-evolutionary coin is much less known and the ideas are presented here in a systematic way probably for the first time, although some material was already available in the form of technical reports. Chapter Eight is also rather unusual in that its main theme, parallelism, is normally dealt with superficially and only with implementation purposes in mind. In other words, the use of parallelism is seen mainly from a performance-gain point of view. This aspect is obviously important and is discussed here but we submit that the natural, intrinsic parallelism of bio-inspired soft computing methods is also worth studying in itself. Nature's way, which is decentralized and distributed, is a rich source of ideas that should not be ignored.

All the chapters contain examples and most of them feature full-sized case studies. We believe that the latter are very useful for practitioners and students alike since they enhance the basic ideas and apply them in a real-world setting. These case studies come from our own work or from the work of our colleagues and cover a broad application spectrum including finance, robotics, medicine, and engineering applications.

Although we deal mainly with software simulation of soft computing systems, we believe that this is a field in which there is scope for special purpose hardware implementations, especially in the case of neural machines and fuzzy systems. Some basic architectural ideas on neural machines are presented in Chapters Four and Eight, where massively parallel and reconfigurable systems are briefly described.

Many people have helped shape the overall framework of the book directly and indirectly. We have greatly benefited from discussions with our colleagues Bastien Chopard, Gianni Degli Antoni, Daniel Mange, Eduardo Sanchez, and Moshe Sipper. We are also indebted to Andrés Pérez, Xin Yao, and Carlos Andrés Peña Reyes for critically reading Chapters Two, Four, and Five, respectively. Andrés Pérez also kindly made several illustrations in Chapter Two available. Our colleagues Dario Floreano and Joseba Urzelai and Carlos Andrés Peña Reyes and Moshe Sipper kindly provided the case studies of Chapters Four and Five, respectively. Finally, we thank Robert Barras for preparing many of the figures appearing in Chapters Two, Four, Six, and Eight. We would also like to express our appreciation to the Springer staff, in particular to Alfred Hofmann, who encouraged us throughout and was instrumental in the whole process, and to Anna Kramer and Ulrike Stricker for their fine editing work. We acknowledge our universities, as well as Genetica S.r.l. in Milan, for providing a supportive work environment during the writing stage. Lastly, we would like to thank our wives Célia and Anne for their love, understanding, and patience.

June, 2001
Andrea G. B. Tettamanzi
Marco Tomassini

# Contents

Chapter 1

# Evolutionary Algorithms

## 1.1 Introduction

EVOLUTIONARY algorithms (EAs) are a broad class of stochastic optimization algorithms, inspired by biology and in particular by those biological processes that allow populations of organisms to adapt to their surrounding environment: genetic inheritance and survival of the fittest. These concepts were introduced in the 19th century by Charles Darwin [50] and are still today widely acknowledged as valid, even though complemented with further details [52].

The first proposals in that direction date back to the mid-1960s, when John Holland, of the University of Michigan, introduced genetic algorithms (GAs) [97], Lawrence Fogel and his colleagues, of the University of California in San Diego, started their experiments on evolutionary programming, [71] and Ingo Rechenberg, of the Technical University of Berlin, independently began to work on evolution strategies [185]. Their pioneering work eventually gave rise to a broad class of optimization methods particularly well suited for hard problems where little is known about the underlying search space. The last development of this research thread is so-called genetic programming, introduced by John Koza, of Stanford University [117] at the beginning of the 1990s.

Recent texts of reference and synthesis in the field of evolutionary algorithms are [143, 12].

An evolutionary algorithm maintains a population of candidate solutions for the problem at hand, and makes it evolve by iteratively applying a (usually quite small) set of stochastic operators, known as *mutation*, *recombination*, and *selection*.

Mutation randomly perturbs a candidate solution; recombination decomposes two distinct solutions and then randomly mixes their parts to form novel solutions; and selection replicates the most successful solutions found in a population at a rate proportional to their relative quality.

The initial population may be either a random sample of the solution space or may be seeded with solutions found by simple local search procedures, if these are available.

The resulting process tends to find, given enough time, globally optimal solutions to the problem much in the same way as in nature populations of organisms tend to adapt to their surrounding environment.

## 1.2 Genetic Algorithms

We start by providing a qualitative description of binary-coded evolutionary algorithms, using two simple examples to present the main elements of these techniques.

### 1.2.1 The Metaphor

Essentially, evolutionary algorithms make use of a metaphor whereby an optimization problem takes the place of the environment; feasible solutions are viewed as individuals living in that environment and an individual's degree of adaptation to its surrounding environment is the counterpart of the objective function evaluated on a feasible solution. In the same way, a set of feasible solutions takes the place of a population of organisms. This optimization setting of evolutionary algorithms is useful in applications, but alternative views related to decision theory and machine learning have been proposed [97]. A further interpretation comes from *Artificial Life* quarters, where evolutionary algorithms are seen as artificial counterparts of natural evolution [121].

In evolutionary algorithms selection operates on computer data structures and, in time, their functionalities evolve in a way substantially analogous to how populations of living organisms evolve in a natural setting.

Although the computer model introduces sharp simplifications with respect to the real biological mechanisms, evolutionary algorithms have proved capable of making surprisingly complex and interesting structures emerge. Each structure, or *individual*, may be viewed as a representation, according to an appropriate encoding, of a particular solution to a problem, of a strategy to play a game, of a picture, or even of a simple computer program.

### 1.2.2 Representation

In genetic algorithms, individuals are just strings of binary digits. As computer memory is made up of an array of bits, anything that can be stored in a computer can also be encoded for by a bit string of sufficient length. In a sense, representing solutions to a problem as bit strings is the most general encoding that can be thought of.

### 1.2.3 The Evolutionary Cycle

An evolutionary algorithm starts with a population of randomly generated individuals, although it is also possible to use a previously saved population, or a population of individuals encoding for solutions provided by a human expert or by another heuristic algorithm. In the case of genetic algorithms the initial population will be made up of random bit strings.

Once an initial population has been created, an evolutionary algorithm enters a loop. At the end of each iteration a new population will have been created by applying a certain number of stochastic operators to the previous population. One such iteration is referred to as a *generation*.

The first operator to be applied is *selection*. Its aim is to simulate the Darwinian law of "survival of the fittest". In genetic algorithms, this law is enforced by so-called fitness proportionate selection: in order to create a new intermediate population of $n$ "parents", $n$ independent extractions of an individual from the old population are performed, where the probability of each individual being extracted is linearly proportional to its fitness. Therefore, above average individuals will expectedly have more copies in the new population, while below average individuals will risk extinction.

Once the population of parents, that is of individuals that have been selected for reproduction, has been extracted, the individuals for the next generation will be produced through the application of a number of reproduction operators, which can involve just one parent (thus simulating asexual reproduction), in which case we speak of mutation, or more parents (thus simulating sexual reproduction), in which case we speak of recombination. In genetic algorithms two reproduction operators are used: crossover and mutation.

To apply crossover, couples are formed with all parent individuals; then, with a certain probability, called crossover rate $p_{\text{cross}}$, each couple actually undergoes crossover: the two bit strings are cut at the same random position and the second halves are swapped between the two individuals, thus yielding two novel individuals, each containing characters from both parents.

After crossover, all individuals undergo mutation. The purpose of mutation is to simulate the effect of transcription errors that can happen with a very low probability ($p_{\text{mut}}$) when a chromosome is duplicated. This is accomplished by flipping each bit in every individual with a very small probability, called mutation rate. In other words, each "0" has a small probability of being turned into a "1" and vice versa.

In principle, the above-described loop is infinite, but it can be stopped when a given termination condition specified by the user is met. Examples of termination conditions are:

- a pre-determined number of generations or time has elapsed;
- a satisfactory solution has been found;
- no improvement in solution quality has taken place for a pre-determined number of generations.

All of the above termination conditions are acceptable under some assumptions relevant to the context the evolutionary algorithm is used in.

The evolutionary cycle can be summarized by the following pseudo-code:

*generation* = 0
Seed Population
**while not** *termination condition* **do**
  *generation* = *generation* + 1
  Calculate Fitness
  Selection
  Crossover($p_{\text{cross}}$)
  Mutation($p_{\text{mut}}$)
**end while**

### 1.2.4 A First Example

An example will illustrate the workings of genetic algorithms and show how a few simple concepts borrowed from natural evolution can give rise to a powerful optimization technique. The following example is based on the MAXONE problem.

Suppose that we want to maximize the number of ones in a string of $l$ binary digits. At first sight this might look like a trivial problem, just because we know the solution in advance: a string of $l$ ones. However, if we imagine being faced with $l$ yes/no answers to an equal number of difficult questions, the problem of maximizing the number of correct answers becomes less straightforward. But then, we can transform that problem into our MAXONE problem simply by assuming that, for each question, the right answer, be it yes or no, is encoded by 1 and the wrong one is encoded by 0.

The fitness of a candidate solution to the MAXONE problem is the number of ones in its genetic code, the string of $l$ binary digits.

We start with a population of $n$ random strings. Suppose that $l = 10$ and $n = 6$: we toss a fair coin 60 times and we get the following initial population:

$$\begin{array}{ll} s_1 = 1111010101 & f(s_1) = 7 \\ s_2 = 0111000101 & f(s_2) = 5 \\ s_3 = 1110110101 & f(s_3) = 7 \\ s_4 = 0100010011 & f(s_4) = 4 \\ s_5 = 1110111101 & f(s_5) = 8 \\ s_6 = 0100110000 & f(s_6) = 3 \end{array} \tag{1.1}$$

where $f$ is the fitness function, which associates with every binary string $s$ its fitness $f(s)$.

Next we apply fitness proportionate selection with the roulette wheel method: we sum up the fitness of individuals in the population, getting 34. We equate the full circumference of a roulette wheel to the total fitness, 34, and we divide it into sectors proportional to each individual's fitness, then we simulate throwing a ball into it. Therefore, when we spin the wheel, the ball will have a $\frac{7}{34} = 0.2059$ probability of stopping in string $s_1$'s sector, and only a $\frac{3}{34} = 0.0882$ probability of stopping in string $s_6$'s sector.

We repeat the extraction using this method six times, that is as many times as the individuals we need to complete our parent population. Suppose that, after performing selection, we get the following population:

$$\begin{aligned} s'_1 &= 1111010101 \quad (s_1) \\ s'_2 &= 1110110101 \quad (s_3) \\ s'_3 &= 1110111101 \quad (s_5) \\ s'_4 &= 0111000101 \quad (s_2) \\ s'_5 &= 0100010011 \quad (s_4) \\ s'_6 &= 1110111101 \quad (s_5) \end{aligned} \tag{1.2}$$

We note that string $s_5$ was extracted twice, while string $s_6$ was never extracted, thus being replaced by a copy of string $s_5$.

Next we mate strings for crossover. Since the strings in the parent population have been extracted in a random order, we can just mate $s'_1$ with $s'_2$, $s'_3$ with $s'_4$ and so on. For each couple thus formed, we decide according to crossover probability (for instance 0.6) whether to actually perform crossover or not. Suppose that we decide to actually perform crossover only for couples $(s'_1, s'_2)$ and $(s'_5, s'_6)$. For each couple, we randomly extract a crossover point, for instance 2 for the first couple and 5 for the second couple.

Therefore, for couple $(s'_1, s'_2)$, we will have

$$\begin{aligned} s'_1 &= 11 \cdot 11010101 \\ s'_2 &= 11 \cdot 10110101 \end{aligned} \tag{1.3}$$

before crossover, and

$$\begin{aligned} s''_1 &= 11 \cdot 10110101 \\ s''_2 &= 11 \cdot 11010101 \end{aligned} \tag{1.4}$$

after crossover. We notice that in this particular case no new genetic material is produced, since the two offspring are equal to their parents.

For couple $(s'_5, s'_6)$, we will have

$$\begin{aligned} s'_5 &= 01000 \cdot 10011 \\ s'_6 &= 11101 \cdot 11101 \end{aligned} \tag{1.5}$$

before crossover, and

$$\begin{aligned} s''_5 &= 01000 \cdot 11101 \\ s''_6 &= 11101 \cdot 10011 \end{aligned} \tag{1.6}$$

after crossover: this time, individuals $s''_5$ and $s''_6$ are novel, each retaining characters from both parents.

The final step to produce the population for the next generation is to apply random mutation: for each bit that we are to copy to the new population we allow a small probability of error (for instance 0.1). Since we have 60 bits overall to transcribe, we expect that on average 6 of them will end up being flipped, for instance according to the following pattern (bits that will be flipped are marked with a bar on top):

$$\begin{aligned} s''_1 &= 11101\bar{1}0101 \\ s''_2 &= 1111\bar{0}1010\bar{1} \\ s''_3 &= 11101\bar{1}11\bar{0}1 \\ s''_4 &= 0111000101 \\ s''_5 &= 0100011101 \\ s''_6 &= 11101100\bar{1}1 \end{aligned} \tag{1.7}$$

Mutations do not have to be evenly distributed over the individuals. In this example, individuals $s''_2$ and $s''_3$ were particularly "unlucky", while individuals $s''_4$ and $s''_5$ passed this stage untouched. If we carefully look a little more at the example, we can observe that of six transcription errors, four make a "1" become a "0", thus worsening the string in which they occur: this should come to no surprise, since as a population adapts to its environment, "good" genes tend to be more frequent than "bad" genes; therefore, transcription errors, which blindly strike at random, will be much likely to disrupt good genes and only occasionally will they introduce a fortuitous improvement.

After applying mutation, we end up with the following new population:

$$\begin{aligned} s'''_1 &= 1110100101 & f(s'''_1) &= 6 \\ s'''_2 &= 1111110100 & f(s'''_2) &= 7 \\ s'''_3 &= 1110101111 & f(s'''_3) &= 8 \\ s'''_4 &= 0111000101 & f(s'''_4) &= 5 \\ s'''_5 &= 0100011101 & f(s'''_5) &= 5 \\ s'''_6 &= 1110110001 & f(s'''_6) &= 6 \end{aligned} \tag{1.8}$$

In one generation, the total fitness of the population passed from 34 to 37, thus improving it by almost 9%. At this point, we go through the same process all over again, getting populations for generation $2, 3, \ldots$, until a stopping criterion is met.

### 1.2.5 A Second Example

In this section we present a second example of the operation of the genetic algorithm, this time involving real function optimization. This should bring things closer to the actual use of GAs, although the problem is purely of illustrative value and can in fact be solved by hand.

The non-constrained function minimization problem can be cast as follows. Given a function $f(\mathbf{x})$ and a domain $D \in \mathbb{R}^n$, find $\mathbf{x}^*$ such that:

$$f(\mathbf{x}^*) = \min\{f(\mathbf{x}) \mid \forall x \in D\}$$

where $\mathbf{x} = (x_1, x_2, \ldots, x_n)^T$.

Let us consider the following function (see Figure 1.1):

$$f(x) = - \mid x \sin(\sqrt{\mid x \mid}) \mid + C$$

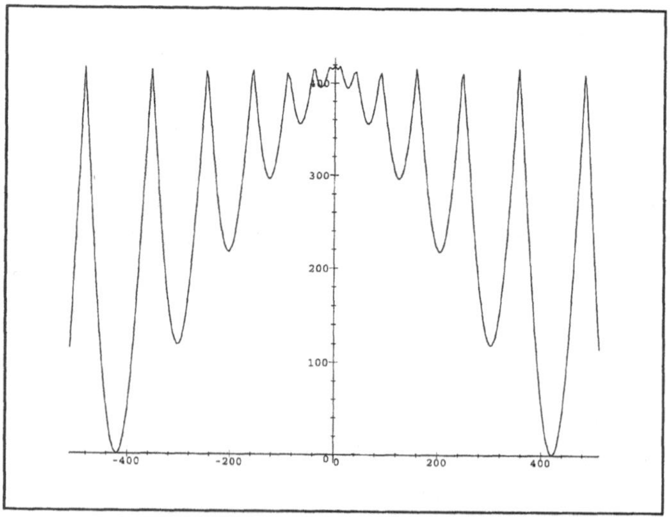


Figure 1.1 Graph of $f(x), x \in [-512, 512]$.

The problem is to find $x^*$ in the interval $[-512, 512]$ which minimizes $f$. Since $f(x)$ is symmetric, studying it in the positive portion of the $x$ axis will suffice.

Let us examine in turn the components of the genetic algorithm for solving the given problem.

The initial population will be formed by 50 randomly chosen trial points in the interval $[0, 512]$. Therefore, one individual is a value of the real variable $x$.

A binary string will be used to represent the values of $x$. However, this time a decoding process from bit strings to real values will be needed, contrary to the previous example. The length of the string will be a function of the required precision; the longer the string the better the precision. For example, if each point $x$ is represented by 10 bits then 1024 different values are available for covering the interval $[0, 512]$ with 1024 points, which gives a granularity of 0.5 for $x$ i.e., the genetic algorithm will be able to sample points no less than 0.5 apart from each other.

The strings 0000000000 and 1111111111 will represent respectively the lower and upper bounds of the search interval. Any other 10-bit string will be mapped to an interior point. In order to map the binary string to a real number, the string is first converted to a decimal number and then to the corresponding real $x$. Note that our use of 10-bit strings is only for illustrative purposes; in real applications, finer granularities and therefore longer strings are often needed.

The fitness of each sample point $x$ is simply the value of the function at that point. Since we want to minimize $f$, the lower the value of $f(x)$ the fitter is $x$.

As in the MAXONE example, we apply fitness proportionate selection with the roulette wheel method. We saw that with this selection method fitter members are more likely to be reproduced; furthemore, strings can be selected more than once. As before, once the new population has been produced, strings are paired at random and recombined through crossover and the offspring replace their parents in the population of the next generation. After crossover, mutation is applied to population members. To measure the quality of our solutions we record both the average population fitness and the fitness of the best individual at a given generation. As an example consider the following table, showing the results of a particular evolutionary run.

| Generation | Best | Average |
|---|---|---|
| 0 | 1.0430 | 268.70 |
| 3 | 1.0430 | 78.61 |
| 9 | 0.00179 | 32.71 |
| 18 | 0.00179 | 14.32 |
| 26 | 0.00179 | 5.83 |
| 36 | 0.00179 | 2.72 |
| 50 | 0.00179 | 1.77 |
| 69 | 0.00179 | 0.15 |

As generation 0 consists of randomly generated individuals, we find, as expected, that both the average and the best fitness values are low. We observe that fairly rapid improvement ensues, with the minimum already found at generation 9. However, the average population fitness continues to improve until the population becomes homogenous and fitness values level off. This behavior is in fact characteristic of evolutionary algorithms in general. Note that in our simple example the probability of getting "stuck" in a local minimum is practically zero. In harder problems, a compromise must be reached between *exploitation* of "good" regions of the search space, (i.e., local improvement), and further *exploration* of this space, in order to find possibly better extrema points.

One final remark is in order; genetic algorithms are stochastic, thus their performance varies between different runs (unless the same random number generator with the same seed is used). Thus, the average performance taken over several runs is a more useful indicator of their behavior than a single run.

The problem presented above is an easy one for GAs, as well as for any other optimization method. GAs have been shown to be effective in solving hard mathematical optimization problems, involving multimodal functions of several variables [152].

## 1.3 Theoretical Background of Genetic Algorithms

THIS SECTION gives an account of the theoretical background of simple evolutionary algorithms with binary representation, the genetic algorithms.

### 1.3.1 Notation

The most general formulation of an optimization problem [163] is as follows:

$$\min c(s) \tag{1.9}$$

subject to the constraint

$$s \in S. \tag{1.10}$$

$S$ is called the *feasible set* of the problem domain, $c$ is a *cost function* and $s \in S$ is called a *feasible solution* to the problem at hand.

Let $\Gamma$ be the space of genotypes. A genotype is an arbitrary data structure that encodes a certain solution to the problem. The space of genotypes is just the collection of all such data structures. We shall assume in the sequel that a genotype can only encode a feasible solution; this hypothesis can be relaxed in various ways that will be dealt with in Section 1.7.3.

Which solution a given genotype encodes is established by a function $M: \Gamma \to \Phi$, where $\Phi \subseteq S$ is the space of phenotypes, for analogy with natural genetics, in which the phenotype is the set of characters shown by an actual realization of a genotype in a living organism. Defining the function $M$ amounts to choosing a particular encoding for solutions. Of course, different genotypes may encode for the same solution, or phenotype, but the reverse does not hold.

The concept that relates to the cost function $c(\cdot)$ is that of *fitness*. The fitness function is a function $f: \Gamma \to [0, +\infty)$, that depends on the cost of a solution through some appropriate transformation function $F$,

$$f(\gamma) = F[c(M(\gamma))], \tag{1.11}$$

such that a larger fitness value corresponds to a better (lower cost) solution: for all $\gamma, \kappa \in \Gamma$,

$$f(\gamma) > f(\kappa) \quad \text{if and only if} \quad c(M(\gamma)) < c(M(\kappa)). \tag{1.12}$$

A population is a collection of individuals, each having its genotype and, as a consequence, the corresponding phenotype. Let us denote by $\Gamma^*$ the space of all populations consisting of any number of individuals. More than one individual in a population can share the same genotype and what distinguishes two populations of equal size is only the number of individuals in which each genotype occurs.

Therefore, a convenient way to think of a population is as a multiset, or bag, of genotypes, that is, the equivalence class of $n$-tuples identical up to a permutation of their elements.

Every population $x \in \Gamma^*$ is in a 1-to-1 correspondence with a share function $q_x: \Gamma \to [0,1]$ giving the fraction $q_x(\gamma)$ of individuals in $x$ that have genotype $\gamma$. The size of population $x$ can be conveniently denoted as $\|x\|$.

### 1.3.2 Main Theoretical Results

#### The Schema Theorem

An important concept for the analysis of genetic algorithms is that of *schema*.

A *schema* is a subset $S \subseteq \Gamma$ represented by a template string consisting of $l$ symbols in $\{0, 1, \star\}$, where "$\star$" plays the role of "wild card": a schema thus contains all strings that match its template string in all positions that are not marked by the "$\star$" symbol.

There are in $\Gamma$ exactly $3^l$ distinct schemata. Each schema induces a bipartition of $\Phi$ and $M(S) \subseteq \Phi$ is called the *hyperplane* defined by $S$.

The *order* $o(S)$ of schema $S$ is defined as the number of fixed positions (0 or 1) in the template string representing it. The cardinality of schema $S$ is bound to its order by the relationship $\|S\| = 2^{l-o(S)}$. On the other hand, a string of length $l$ is matched by $2^l$ schemata.

The *absolute* fitness of schema $S$ is

$$f(S) \equiv \frac{1}{\|S\|} \sum_{\gamma \in S} f(\gamma). \tag{1.13}$$

This quantity represents the expected fitness of an individual randomly extracted with uniform probability from $S$.

The *relative* fitness of schema $S$ with respect to population $x \in \Gamma^*$ is

$$f_x(S) \equiv \frac{1}{q_x(S)} \sum_{\gamma \in S} q_x(\gamma) f(\gamma). \tag{1.14}$$

This quantity represents the expected fitness of an individual randomly extracted from population $x$, given that it belongs to schema $S$.

The *defining length* $\delta(S)$ of schema $S$ is the distance between the first and last fixed position in its template string. The defining length can be interpreted as a measure of information "compactness" in a schema.

Let $\{X_t\}_{t=0,1,\ldots}$ be the sequence of populations generated by the genetic algorithm at generations $t = 0, 1, \ldots$; assuming constant the ratio

$$c = \frac{f_{X_t}(S) - f(X_t)}{f(X_t)}, \quad t = 0, 1, \ldots, \tag{1.15}$$

we have that

$$E[q_{X_t}(S)|X_0] \geq q_{X_0}(S)(1+c)^t \left(1 - p_{\text{cross}}\frac{\delta(S)}{l-1} - o(S)p_{\text{mut}}\right)^t, \tag{1.16}$$

where $p_{\text{cross}}$ and $p_{\text{mut}}$ are, respectively, the crossover and mutation rates.

In other words, it is expected that short, low-order, above-average schemata get an exponentially increasing number of instances in subsequent generations.

This result, known as the Schema Theorem, can be proved as follows. We calculate first of all the conditional expectation $E[q_{X_t}(S)|X_{t-1}]$ keeping into account the cumulative effects of crossover and mutation. Using fitness proportionate selection, we get

$$E[q_{X_t}(S)|X_{t-1}] \geq q_{X_{t-1}}(S)\frac{f_{X_{t-1}}}{f(X_{t-1})}P_{\text{surv}}[S] = q_{X_{t-1}}(S)(1+c)P_{\text{surv}}[S], \tag{1.17}$$

where $P_{\text{surv}}[S]$ is the probability for the fixed positions of schema $S$ of not being touched by crossover and mutation. Clearly, $P_{\text{surv}}[S] \geq P_{sc}[S]P_{sm}[S]$, where $P_{sc}[S]$ is the probability that $S$ survives crossover and $P_{sm}[S]$ is the probability that $S$ survives mutation, because it could happen that a chance mutation restores a part of $S$ corrupted by crossover.

Therefore, $P_{sc}[S]$ and $P_{sm}[S]$ have to be calculated. It should be clear that the defining length of a schema plays a significant role in its probability of survival: the crossover point being uniformly chosen among $l-1$ possible points, the probability that the chosen point does not split the fixed part of a schema is given by $1 - \frac{\delta(S)}{l-1}$; now, there is a probability $p_{\text{cross}}$ that an individual undergoes crossover, whence it can be written

$$P_{sc}[S] = 1 - p_{\text{cross}}\frac{\delta(S)}{l-1}. \tag{1.18}$$

As for mutation, the probability that each single position in a genotype is altered is $p_{\text{mut}}$ and, given that mutation in each position is independent of mutation in the others, the probability that no fixed position is altered is

$$P_{sm}[S] = (1 - p_{\text{mut}})^{o(S)} \approx 1 - o(S)p_{\text{mut}}, \tag{1.19}$$

since $p_{\text{mut}} \ll 1$. This makes it possible to write

$$E[q_{X_t}(S)|X_{t-1}] \geq q_{X_{t-1}}(S)(1+c)\left(1 - p_{\text{cross}}\frac{\delta(S)}{l-1} - o(S)p_{\text{mut}}\right). \tag{1.20}$$

Substituting $q_{X_{t-1}}(S)$ with $E[q_{X_{t-1}}(S)|X_{t-2}]$ and iterating the reasoning until we get to $q_{X_0}(S)$ yields the thesis.

### The Building Blocks Hypothesis

A consequence of the schema theorem is the so-called building blocks hypothesis [143], which states that

> An evolutionary algorithm seeks near-optimal performance through the juxtaposition of short, low-order, high performance schemata: the building blocks.

When the building block hypothesis does not hold we have *deception.* The simplest case of deception happens when, for some schema $S$, $\gamma^* \in S$ but $f(S) < f(\bar{S})$, where $\gamma^*$ is the optimal genotype and $\bar{S}$ is the complement of $S$. In such cases, the genetic algorithm is deceived by schemata that are above average, but that do not lead in the right direction.

Three remedies to deception have been proposed:

- if *a priori* knowledge of the objective function is available, one can use it to come up with a non-deceptive encoding;
- one can introduce a new inversion operator which makes the semantics of genes non-positional;
- genotypes might underspecify or overspecify a solution, this being the solution adopted by Goldberg's messy genetic algorithms [77].

### Convergence in Probability

It has been proved [1, 51] that, under certain rather mild assumptions, the process $\{X_t\}_{t=0,1,2,\ldots}$ converges in probability to a global optimum.

This is no impressive result: other very inefficient search algorithms, like random search and exhaustive search, enjoy the same property. What convergence in probability means is that, provided we have enough patience, evolution will always find the best solution.

### Rate of Convergence

Although we know that an evolutionary algorithm will always, sooner or later, find the optimal solution for every problem, we still have no clue as to how long we will have to wait.

The general rate of convergence of evolutionary algorithms is still an open question, and a very difficult one at that.

It is reasonable to conjecture that all problems might be characterized by the rate at which a well designed evolutionary algorithm converges to their solution. Evolutionary algorithms might thus give rise to their own problem complexity classes, and these classes might have some relationship with the traditional computational complexity classes of computer science.

While this could be an ambitious program for future research in the field of evolutionary computing, to date, to the authors' knowledge, no significant result has been achieved in this direction.

## 1.4 Introduction to Classifier Systems

CLASSIFIER systems (CS) provide another look at genetic algorithms, one which is not based on optimization concepts but rather on machine learning ideas. Machine learning is a field of artificial intelligence which deals with the question of how to build computer programs that can improve their behaviour by experiencing their environment; that is, they should be able to change the course of calculations as the model accumulates experience. One early and important piece of work in this field was Samuel's checkersplaying program, which is still significant today [197]. There exist many specific learning techniques and we cannot discuss all of them here; the interested reader can consult Mitchell's book for a complete survey of the subject [148]. We will meet several machine learning ideas especially in connection with neural networks. Here we describe classifier sytems, a machine learning technique that has deep connections with genetic algorithms and that will be extended to the fuzzy world in Chapter 5. For an in-depth treatment of the subject see Goldberg's book [76].

Classifier systems are rule-based systems. Rule-based systems belong to the class of *production systems*, a computational class of systems consisting of a set of rules each having a left side that determines the applicability of the rule and a right side that describes what is to be done if the rule is applied. Each rule maps a problem state into a new state, where the process is applied again, or a solution is found. The rules in classifier systems are called *classifiers* and are of the following form:

$$\text{IF } \langle \textit{condition} \rangle \text{ THEN } \langle \textit{action} \rangle$$

which is to be interpreted such that $\langle \textit{action} \rangle$ is executed if $\langle \textit{condition} \rangle$ is true. Unlike classical expert systems, which need access to a substantial domain knowledge base, usually established through human experts and translated into rules, classifier systems may discover new rules through simulated evolution. Furthermore, rules in a classifier systems are processed in parallel as a whole, whereas traditional expert systems only fire one rule at a time in sequential fashion.

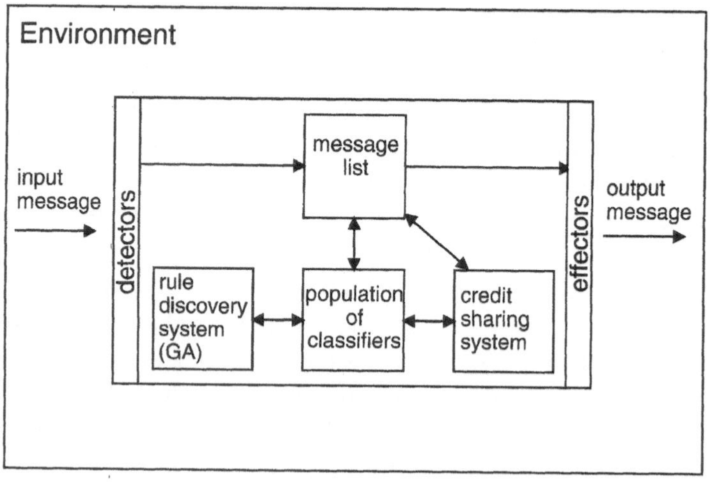


Figure 1.2 Schematic architecture of a learning classifier system.

A classifier system has three main components:

- message and rule system
- credit-assignment system
- genetic algorithm

Classifier systems work roughly in the following way (see also Figure 1.2). A population of rules is encoded as bit strings and evolves over time. The environment sends messages representing example events to the CS which are decoded by the detectors and placed into an internal message list. The classifier system then tries to match messages with one or more classifier conditions. If some conditions match, an action is selected and applied by the effectors through a coding of the corresponding output message. Classifiers are given fitness values, also called *strength*, through a credit-assignment system, and new rules can be discovered in the classifier population by the genetic algorithm. The syntax of a classifier (rule) is of the type:

$$\langle condition \rangle \ : \ \langle message \rangle$$

A *message* is simply a finite string over some alphabet, e.g. the binary alphabet $\{0,1\}^k$. The condition part of the rule can also be coded with the same alphabet plus a "wild card" symbol $\star$ (see Section 1.3.2) that matches either a 0 or a 1. Thus, for example, both the messages 0011 and 0101 would match the condition $0{\star}{\star}1$. When a message matches a classifier condition the classifier may place its own message part into the message list. Classifiers are all of the same length.

When classifiers are activated by a message, a market-like mechanism is employed to choose which one will win and be able to post its message into the list and how the partial worth of other classifiers in the activation chain is to be acknowledged. Several solutions have been proposed for sharing the credit among classifiers, all of them being relatively involved. For the sake of simplicity, it suffices to say here that the classifier worth, called its *bid*, is proportional to its strength and that a reward system is set up such that successful classifiers get a positive reward that increases their current strength. However, the successful classifier recognizes its debt by sharing its bid in some way among those classifiers that sent the messages that matched the bidding classifier's condition. More details on the credit-sharing problem may be found in [76] where the original Holland's "bucket brigade" algorithm is described as well as more recent variations.

The genetic algorithm part of the classifier system allows it to generate new, possibly superior, classifiers. New rules are built by recombining and mutating the current rules in the population. Crossover is identical to the one used in GAs for search and optimization purposes. Mutation is also the same but, since the wild card symbol is also permitted, when a symbol is to be mutated it can change to one of the other two with equal probability. Another difference with the classical GA is that in machine learning the whole population of classifiers co-evolves and learns collectively. Replacing the whole population at once after the selection phase would disrupt the learning process. In practice, individuals are replaced more gently and taking into account their degree of similarity.

The above schematic presentation of classifier systems corresponds to the original model proposed by Holland (see for instance [98] and [76] for an historical perspective on the development of CS). A more recent version of classifier systems is the one presented by Wilson [246], called the XCS classifier system. XCS differs from the standard CS model in various respects, especially in the way actions are selected after a set of classifiers that match a given message has been formed. The apportionment of credit algorithm is also different and is based on reinforcement learning ideas (see Chapters 2 and 4).

For the sake of completeness, it is to be noted that classifier systems are not the only evolutionary approach to machine learning. In the so-called Pitt approach, instead of having a population of co-evolving rules that are to be considered as an integrated set, each individual represents a set of rules which are complete solutions to the problem (see [143]). Genetic programming, which is the subject of the next section, can also be considered as an evolutionary machine learning technique.

## 1.5 Genetic Programming

GENETIC programming (GP) is a new evolutionary approach which extends the genetic model of learning to the space of programs. It is a major variation of genetic algorithms in which the evolving individuals are themselves computer programs instead of fixed length strings from a limited alphabet of symbols. Genetic programming is a form of *program induction* that can be used to automatically discover programs that solve or approximately solve a given task. The present form of GP is principally due to J. Koza[117].

Individual programs in GP might be expressed in principle in any current programming language. However, the syntax of most languages is such that GP operators would create a large percentage of syntactically incorrect programs. For this reason, Koza chose a syntax in prefix form analogous to LISP and a restricted language with an appropriate number of variables, constants and operators defined to fit the problem to be solved. In this way syntax constraints are respected and the program search space is limited. The restricted language is formed by a user-defined *function set* $F$ and *terminal set* $T$. The functions chosen are those that are *a priori* believed to be useful for the problem at hand, and the terminals are usually either variables or constants. In addition, each function in the function set must be able to accept as arguments any other function return value and any data type in the terminal set $T$, a property that is called *syntactic closure*. Thus, the space of possible programs is constituted by the set of all possible compositions of functions that can be recursively formed from the elements of $F$ and $T$.

As an example, suppose that we are dealing with simple arithmetic expressions in four variables. In this case, suitable function and terminal sets might be defined as:

$$F = \{+, -, *, /\}$$

and

$$T = \{\mathtt{A}, \mathtt{B}, \mathtt{C}, \mathtt{D}\}$$

and the following are legal programs: `(+ (* A B) (/ C D))`, and `(* (- (+ A C) B) A)`.

It is important to note that GP does not need to be implemented in the LISP language (though this was the original implementation). Any language that can represent programs internally as parse trees is adequate. Thus, most GP packages today are written in C, C++ or Java rather than LISP. GP representation is not restricted to trees however. Other program representations have been proposed such as linear and graph [18].

For the sake of simplicity and generality, we will depict programs as trees with ordered branches in which the internal nodes are functions and the leaves are the terminals of the problem. Thus, the examples given above would give rise to the trees in Figure 1.3.

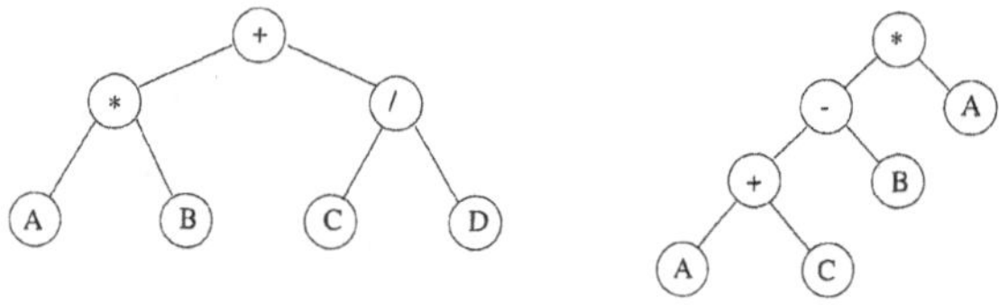


Figure 1.3 Two GP trees corresponding to the LISP expressions in the text.

Evolution in GP is similar to GAs, except that different individual representation and genetic operators are used. Once suitable functions and terminals have been determined for the problem at hand, an initial random population of trees (programs) is constructed. From there on the population evolves as with a GA where fitness is assigned after actual execution of the program (individual) and with genetic operators adapted to the tree representation. Fitness calculation is a bit different for programs. In GP we would like to discover a program that satisfies a given number $N$ of predefined input/output relations: these are called the *fitness cases*. For a given program $p_i$ its fitness $f_j(p_i)$ on the $j$-th fitness case represents the difference between the output $g_j$ produced by the program and the correct answer $G_j$ for that case. The total fitness $f(p_i)$ is the sum over all $N$ fitness cases of some norm of the cumulated difference:

$$f(p_i) = \sum_{k=1}^{N} \| g_k - G_k \| . \tag{1.21}$$

Obviously, a better program will have a lower fitness under this definition, and a perfect one will score 0 fitness.

The crossover operation starts by selecting a random crossover point in each parent tree and then exchanging the sub-trees, giving rise to two offspring trees, as shown in Figure 1.4. The crossover site is usually chosen with non-uniform probability, in order to favor internal nodes with respect to leaves. Mutation is implemented by randomly removing a subtree at a selected point and replacing it with a randomly generated subtree, although this operator is seldom used.

One problematic step in GP is the choice of the appropriate language for a given problem. In general, the problem itself suggests a reasonable set of functions and terminals but this is not always the case. Although experimental evidence has shown that good results can be obtained with slightly different choices of $F$ and $T$, it is clear that the choice of language has an influence on how hard the problem will be to solve with GP. For the time being, there is no guideline for estimating this dependence nor for choosing suitable terminal and function sets. Self-adaptation or co-evolution (see Section 1.7.5) of functions and terminals might help in finding good language building blocks.

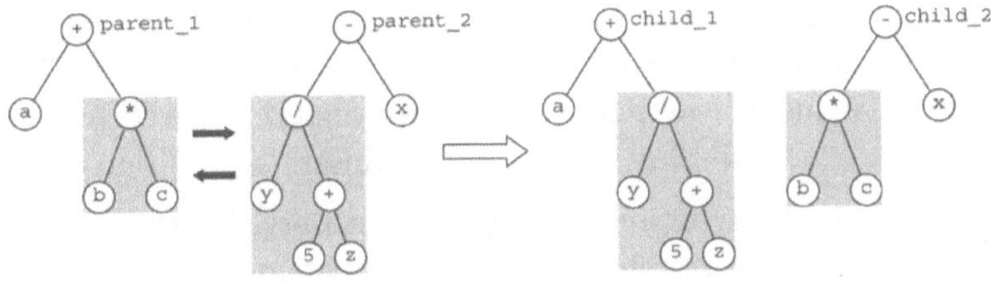


Figure 1.4 Example of crossover of two genetic programs.

Another controversial issue has to do with the size of the GP trees. The depth of the trees can in principle increase without limits under the influence of crossover, a phenomenon that goes under the name of "bloating". The increase in size is often accompanied by a stagnant population fitness. Most GP systems have a parameter that prevents trees from becoming too deep, thus filling all the available memory and requiring longer evaluation times. To further avoid bloating, a common approach is to introduce a size-penalty term into the fitness expression, possibly in a self-adapting way. There is still some debate among practitioners in the field as to whether one should let the trees breed and grow until the maximum depth or whether to edit and simplify them along the way in order to obtain shorter programs. The argument for larger trees is that the often redundant genetic material has a richer set of breeding possibilities and may lead to increased diversity in successive populations. On the other hand, the use of parsimony through Minimum Description Length principles or size penalties may give rise to compact and efficient solutions in some cases [263]. The issue is difficult to settle due to our currently limited knowledge about the dynamics of the evolution of program populations.

Plain GP works well for problems that are not too complex and that give rise to relatively short programs. To extend GP to more complex problems some hierarchical principle has to be introduced. In any problem-solving activity hierarchical considerations are needed to produce economically viable solutions. This is true in classical top-down design where some form of divide-and-conquer strategy is routinely used to decompose the problem into manageable subproblems. The same considerations are also useful when working bottom-up, as in artificial evolutionary methods. It has been observed by several researchers that during evolution some subtrees appear repeatedly within the population as parts of successful individuals. Those trees that seem to perform a useful function might be identified, encapsulated into modules, and reused as single units in the evolutionary process. Methods for automatically identifying and extracting useful modules within GP have been discussed by Koza under the name of Automatically Defined Functions (ADF) ([118]), by Angeline and Kinnear in [59], and by Rosca in [189].

GP is particularly useful for program discovery, i.e. the induction of programs that correctly solve a given problem with the assumption that the form of the program is unknown and that only the desired behavior is given, e.g. by specifying input-output relations. Genetic programming has been successfully applied to a wide variety of problems from many fields, described in [117, 118, 18] and, more recently, in [5, 59]. In conclusion, GP has been empirically shown to be quite a powerful automatic or semi-automatic program-induction and machine learning methodology.

## 1.6 Evolution Strategies and Evolutionary Programming

THIS SECTION provides an overview of evolution strategies and evolutionary programming along with some relevant theoretical results.

### 1.6.1 Evolution Strategies

Evolution strategies [202, 185, 203] approach function optimization problems in the $l$-dimensional real space by exploiting a real encoding of the objective function parameters.

Phenotypes are $l$-dimensional vectors, i.e. $\Phi = \mathbb{R}^l$. A genotype is made up of the same vector as the associated phenotype, plus up to $l$ variances $c_{ii} = \sigma_i^2$, with $i = 1, \ldots, l$, and up to $l(l-1)/2$ covariances $c_{ij}$, with $i, j = 1, \ldots, l$, of the $l$-dimensional normal joint distribution having vector $\mathbf{0}$ as its expectation and density function, for all $\mathbf{z} \in \mathbb{R}^l$,

$$p(\mathbf{z}) = \sqrt{\frac{\det \mathbf{C}^{-1}}{(2\pi)^l}} e^{-\frac{1}{2}\mathbf{z}^T\mathbf{C}^{-1}\mathbf{z}}, \tag{1.22}$$

where $\mathbf{C} = (c_{ij})$ is the variance/covariance matrix. The choice of a normal distribution, that will be used to perturb the genotypes, is obviously arbitrary. Overall, an individual will contain $k \leq l(l+1)/2$ parameters relevant to the "strategy" besides the $l$ parameters relevant to the object problem, whence in general $\Gamma = \mathbb{R}^{k+l}$; often, however, only variances are considered, whereas sometimes it is sufficient to consider one variance for all the object problem parameters.

The fitness of an individual $\gamma \in \Gamma$ is obtained by scaling the objective function to make it positive and, sometimes, by adding a random noise described as a random variable $W$,

$$f(\gamma) = F\{c[M(\gamma)], W\}. \tag{1.23}$$

**Mutation**

In its most general form, the mutation operator perturbs a genotype $\gamma = (\mathbf{z}, \mathbf{C})$ by first randomly modifying $\mathbf{C}$ and then $\mathbf{z}$ according to the new probability distribution provided by $\mathbf{C}$, thus producing a new individual $\gamma' = (\mathbf{z}', \mathbf{C}')$, where

$$\mathbf{z}' = \mathbf{z} + \mathcal{N}(\mathbf{0}, \mathbf{C}'), \tag{1.24}$$

where $\mathcal{N}(\mathbf{0}, \mathbf{C}')$ denotes a random vector with normal joint distribution with mean $\mathbf{0}$ and variance/covariance matrix $\mathbf{C}'$.

This mutation mechanism allows the algorithm to autonomously evolve the parameters relevant to its strategy while searching for the solution: the resulting evolutionary process has been called *self-adaptation* [204].

In practice, instead of modifying $\mathbf{C}$ directly, contemporary evolution strategies use the decomposition $\mathbf{C} = (\mathbf{ST})^T(\mathbf{ST})$ [193], where $\mathbf{S}$ is the diagonal matrix of standard deviations ($s_{ii} = \sigma_i$), and

$$\mathbf{T} = \prod_{i=1}^{l-1} \prod_{j=i+1}^{l} \mathbf{R}_{ij}(\alpha_{ij}) \tag{1.25}$$

is the product of $l(l-1)/2$ elementary rotation matrices[1] $\mathbf{R}_{ij}$ with angles $\alpha_{ij} \in (0, 2\pi]$. The $l(l-1)/2$ rotation angles and the $l$ standard

[1] An elementary rotation matrix $\mathbf{R}$ with an angle $\alpha$ is obtained from an identity

deviations can be directly used to generate a joint normal vector deviate

$$\Delta \mathbf{z} = \mathbf{T}^T \mathbf{S}^T \mathcal{N}(\mathbf{0}, \mathbf{I}),$$

with the same probability density function as in Equation 1.22.

On the basis of this, mutation in its most general form is a three-stage operation, consisting in

1. perturbing every rotation angle $\alpha$ according to $\alpha' = \alpha + \varphi\mathcal{N}(0,1) \pmod{2\pi}$;

2. updating the standard deviations for each variable by the log-normal self-adaptation method, whereby $\sigma_i' = \sigma_i e^{\tau\mathcal{N}(0,1)}$;

3. applying Equation 1.24 in the form

$$\mathbf{z}' = \mathbf{z} + \mathbf{T}^T(\alpha')\mathbf{S}^T(\sigma')\mathcal{N}(\mathbf{0}, \mathbf{I}).$$

### Recombination

Evolution strategies utilize various recombination mechanisms, which in the simplest case produce one child individual from a couple of parents or, in the global case, can form the new individual by combining all the individuals in the population (orgy).

The most widely used recombination mechanisms are discrete and intermediate recombination: in discrete recombination each component is copied from one of the parents at random; in intermediate recombination the value of each component of the child individual is a linear combination of the corresponding component of all the parents participating in the operation.

It has been observed that the best results are obtained by applying discrete recombination to the object problem parameters and intermediate recombination to the strategy parameters. Furthermore, it has been proved that recombination of the latter is required for self-adaptation phenomena to take place.

---

matrix by replacing four entries, identified by indices $i$ and $j$, as follows:

$$\begin{aligned} r_{ii} = r_{jj} &= \cos\alpha, \\ r_{ij} = -r_{ji} &= -\sin\alpha. \end{aligned}$$

### Selection

Selection in evolution strategies is deterministic, according to two alternative schemes, which define two classes of strategies, $(n, m)$ and $(n + m)$. In $(n, m)$ strategies, from a population of $n$ individuals, $m > n$ offspring are produced and the best $n$ of them are kept for the next generation. The $n$ parents are always discarded to make room for the best offspring. In $(n + m)$ strategies, on the contrary, the best $n$ individuals among the $m$ offspring and the $n$ parents survive into the next generation: an $(n + m)$ strategy never discards the best solutions so-far (elitism), thus guaranteeing a monotone improvement of the population; on the other hand such a strategy has a hard time reacting to problems that change in time and does not support evolution of strategy parameters in a satisfactory way, in particular for small populations. For these reasons $(n, m)$ strategies are nowadays preferred, where experiments point to an optimal fraction $n/m \approx \frac{1}{7}$ [204].

### Theoretical Results

An early theoretical result was the so-called $\frac{1}{5}$-*success rule*, stated by Rechenberg [185], which provides a method for controlling the standard deviation on the basis of the observed frequency of mutations resulting in an improvement of the individuals undergoing them (success):

> The optimal fraction of success over all mutations is $\frac{1}{5}$. If it is greater than $\frac{1}{5}$, increase the standard deviation; if it is less, decrease it.

There are two problems concerning the use of this rule:

- sometimes the success rate remains below $\frac{1}{5}$ even when the standard deviation is decreased to zero;
- the rule gives no suggestion as to how single deviations should be treated individually and therefore does not allow the scaling of the average mutation step along distinct axes of the coordinate systems in a different way.

It follows that this result is really useful only for strategies with one standard deviation and not using recombination and self-adaptation. It has to be said that this rule was never intended to be used with anything other than $(1 + 1)$ evolution strategies.

The general expression for the convergence rate, i.e. the expected rate of improvement for the average fitness in a population $\varphi_t = E[f(X_{t+1})/f(X_t)|X_t]$ for $(n, m)$ and $(n + m)$ strategies using one standard deviation and without recombination and self-adaptation was obtained by Schwefel [203].

A recent result for evolution strategies is convergence in probability for the $(1 + 1)$ strategy proven by Günter Rudolph [192]. His proof can be extended to the general $(n + m)$ case, but not to $(n, m)$ strategies.

### 1.6.2 Evolutionary Programming

Evolutionary programming [70, 71] is an approach to Artificial Intelligence making use of finite states automata. Intelligent behavior requires both the capability of predicting the environment and mechanisms to translate those predictions into reactions appropriate for reaching a goal. At the highest level of generality, the environment is described as a sequence of symbols from a given finite alphabet. The task is therefore to evolve an algorithm operating on a sequence of observed symbols and producing an output symbol so as to maximize its performance with respect to the next environment symbol according to a well-defined reward function.

A finite states automaton [99] is a five-tuple $\langle Q, \Sigma, Z, \delta, \omega \rangle$, where $Q$ is the set of internal states, $\Sigma$ is the input alphabet, that is the environment, $Z$ is the output alphabet, $\delta: \Sigma \times Q \to Q$ is the transition function and $\omega: \Sigma \times Q \to Z$ is the output function.

In evolutionary programming the space of phenotypes $\Phi$ is the set of finite states automata with $Q$, $\Sigma$, and $Z$ fixed according to the problem approached. The space of genotypes $\Gamma$ consists in descriptions of functions $\delta$ and $\omega$, typically in the form of $\|Q\| \times \|Z\|$ tables having a pair $(q, a)$, $q \in Q$, and $a \in Z$, in each cell; more compact encodings are often used when the problem structure makes it possible.

Fitness of automata in a population is calculated by applying them to an observed sequence of symbols. One symbol at a time is fed into each automaton and its output compared with the next symbol in the sequence; the accuracy of prediction is measured according to the reward function. When all the symbols have been read, the fitness of each automaton is given by a function of the single rewards (e.g.

the average per-symbol reward).

New automata are generated by random mutation from each automaton already in the population: typically each parent produces one child. There are five kinds of mutation suggested by the description of an automaton:

1. replacing an output symbol;
2. replacing a state in the transition function definition;
3. adding a new internal state;
4. removing an internal state;
5. changing the initial state.

These operations are subject to constraints on the maximum and minimum number of internal states; mutation operates according to an assigned probability distribution, typically uniform; furthermore the number of mutations of an individual can be in turn governed by a probability law, for example a Poisson distribution.

In its original formulation evolutionary programming does not provide for a recombination operator.

The selection process can be performed according to one of several general techniques, including:

1. the best $n$ solutions are retained to become the parents for the next generation (truncation selection);
2. each individual is compared against $k$ randomly chosen other individuals, and the $n$ individuals with the best win records are chosen (a sort of tournament selection);
3. standard fitness proportionate selection, as in genetic algorithms.

### Theoretical Results

David B. Fogel [68] has proved global convergence in probability for evolutionary programming.

## 1.7 Advanced Topics

THIS SECTION gives an idea of the full range of evolutionary techniques, discussing the main issues relevant to their application in practice.

### 1.7.1 Selection Methods and Reproduction Strategies

#### Exploration vs. Exploitation

The purpose of selection in evolutionary algorithms is to concentrate the use of the available computational resources in promising regions of the search space.

There is a relationship of reciprocity between the aspects of *exploration* and *exploitation* of the search space and, clearly, the stronger the pressure exerted by selection toward a concentration of the computational effort, the smaller the fraction of resources utilized to explore other possibilities. At one extreme, as selective pressure decreases and, as a consequence, the resources employed for exploration increase, an evolutionary algorithm tends to behave just like a raw Monte Carlo method, randomly sampling the space of feasible solutions; at the other extreme, as the selective pressure increases, the evolutionary algorithm degenerates into a local "gradient descent" search method.

These two extremes are in a precise correspondence with two typologies of objective functions: on one side gradient descent methods ensure fast convergence to the global optimum for unimodal objective functions; on the other side, the only algorithm ensuring almost certain convergence to the global optimum for highly irregular objective functions (almost everywhere discontinuous, multimodal, noisy, etc.) is, apart from exhaustive search, which is impractical almost for all problems, random sampling.

Evolutionary algorithms can thus be regarded as a trade-off between these extremes and selection is the instrument to adjust it.

#### Fitness Proportionate Selection

Fitness proportionate selection was derived by Holland as the optimal trade-off between exploration and exploitation using an analogy with the $k$-armed bandit [97].

Despite being rigorously derived and having a deep justification in Decision Theory, fitness proportionate selection has some drawbacks.

For instance, consider two fitness functions $f_1(\gamma)$ and $f_2(\gamma) = f_1(\gamma) + c$, for all $\gamma \in \Gamma$. Since they appear substantially equivalent, one would expect the behavior of an evolutionary algorithm not to change by replacing one with the other; however, if the selection scheme is fitness proportionate this is obviously false.

Another difficulty is represented by so-called *superindividuals*. A superindividual in population $x$ is an individual $\gamma \in \Gamma$ such that $f(\gamma) \gg f(x)$. A superindividual is allocated by fitness proportionate selection a prominent slice of copies in the next generation and, in a few generations, it ends up overwhelming any other genotypes initially in the population, causing convergence. If the superindividual corresponds to the problem's global optimum, this is exactly what is desired, but if otherwise it is associated with a local optimum, this leads to a failure of the algorithm, called *premature* convergence.

The list of problems does not end here. The push toward improvement provided by fitness proportionate selection asymptotically tends to zero as the individuals in a population together approach the optimum. While this in fact reflects what is observed in nature, in the framework of optimization it amounts to an actual drawback.

In order to solve these difficulties two approaches are possible: either appropriately modifying the fitness function or, more simply, resorting to alternative selection schemes.

### Linear Ranking Selection

Linear ranking selection [17] is based on a sorting of individuals by decreasing fitness. The probability for the $i$th individual in the ranking of being extracted is thus defined, for $i = 1, \ldots, n$, as

$$p(i) = \frac{1}{n}\left[\beta - 2(\beta - 1)\frac{i-1}{n-1}\right], \qquad (1.26)$$

where $0 \leq \beta \leq 2$ is a parameter that can be interpreted as the expected sampling rate of the best individual across $n$ independent extractions with re-insertion.

### Local Tournament Selection

Local tournament selection [33], extracts $k$ individuals from the population with uniform probability but *without* re-insertion and makes them play a "tournament", which is won, in the deterministic case, by the fittest individual among the participants. The tournament may be probabilistic as well, in which case the probability for an individual to win it is generally proportional to its fitness.

Selective pressure is directly proportional to the number $k$ of participants in a tournament: for $k = n$, the population size, deterministic local tournament selection degenerates into truncation selection with parameter $\tau = \frac{1}{n}$ (see below), whereas probabilistic local tournament selection degenerates into fitness proportionate selection.

### Truncation Selection

Truncation selection [151], has its inspiration in the science of breeding, a branch of applied statistics, and the main concepts on which it relies are the correlation between parent and offspring and the inheritance coefficient.

As Mühlenbein and Schlierkamp-Voosen point out, there is no major difference between breeding natural organisms and solutions to a problem. A minor difference is that in the latter case it is possible to control the genetic operators of mutation and recombination and modify them in order to get the greatest advantage out of them.

The most interesting aspect of selection for a breeder is the *response to selection* $R$, defined as the difference between the average fitness in two subsequent generations:

$$R_t = f(x_{t+1}) - f(x_t). \tag{1.27}$$

Breeders measure selection through the *selective differential* $S$, defined as the difference between the average fitness of the individuals selected for reproduction and the average fitness of the entire population:

$$S_t = f(\bar{x}_t) - f(x_t), \quad \bar{x}_t \sqsubseteq x_t. \tag{1.28}$$

Truncation selection consists in selecting for reproduction just the best individuals and discarding the rest. The selective differential depends on the proportion $\tau$ of the population that gets selected: the smaller $\tau$, the greater $S_t$.

### Classification of Selection Schemes

A possible taxonomy of selection schemes is the following, whereby a selection scheme can be classified according to at least four independent axes [14]:

- *dynamic – static* depending on whether the selection probabilities depend on the fitness values actually present in the population, varying across generations: fitness proportionate selection is thus static, while linear ranking, local tournament and truncation selection are dynamic;

- *preservative – extinctive,* depending on whether it guarantees to every individual a non-zero probability of being selected: thus truncation and deterministic local tournament selection are extinctive, fitness proportional selection is preservative and linear ranking selection is preservative for $\beta < 2$ and extinctive for $\beta \geq 2$;

- *elitist* – *pure*, depending on whether it guarantees the survival of the best individual unchanged into the next generation.
- *generational* – *steady-state*, depending on whether the set of parents is determined once and remains fixed until a new population of offspring has been produced, or the parents are extracted at different times and their offspring are introduced into the population as they are produced.

### 1.7.2 Specialized Representations and Genetic Operators

The recent trend in the effective application of evolutionary algorithms to real world problems is to abandon general encoding schemes like bit strings and rely on the following key ideas:

- use a data structure as close as possible to the natural representation suggested by the object problem;
- write appropriate genetic operators as needed;
- if possible, ensure that all genotypes correspond to feasible solutions;
- if possible, ensure that genetic operators preserve feasibility.

At the same time, it is advisable to represent solutions in such a way that the "genes" which encode them be as *orthogonal* as possible. By orthogonality here it is meant that the semantics of each gene depends only on its *allele* (or value) and is independent of other genes. When this property does not hold, and there are interactions among genes, especially interactions in which one gene suppresses the expression of another, geneticists speak of *epistasis*.

Although it is difficult to formulate a general recipe that makes it possible, given an object problem, to come up with the best encoding scheme for it, it is often useful to compare one's problem with others that have been successfully solved using evolutionary algorithms in the scientific literature, hoping to find one that is roughly similar.

However, it is possible to give a very coarse classification of problems which can serve as a guide for the choice of the representations that are most likely to work well. A partial attempt to list such big classes follows, along with suggestions for appropriate encodings.

#### Pie Problems

Problems that involve finding a (constrained) weight assignment, which can effectively be represented by a pie chart, may be referred to as pie problems. The solution can be encoded as a vector of integer or floating-point numbers and the decoder can be written in such a way as to ensure that no constraint is violated.

For example, if we are given a finite amount of money and we are to find the best way to invest it on $N$ assets, we have the following constraints:

$$\sum_{i=1}^{N} w_i = 1 \quad \text{and, for all } i, \quad w_i \geq 0, \tag{1.29}$$

where $w_i$ is the relative weight of the amount invested on the $i$th asset with respect to the total. These two constraints are easy to enforce when we encode solutions as an unconstrained array of $N$ integers between 0 and a positive constant $g_{\max}$, simply by defining the decoding function $M : \Gamma \to [0,1]^N$ as follows:

$$w_i = \frac{\gamma_i}{\sum_{j=1}^{N} \gamma_j}, \tag{1.30}$$

where $\mathbf{w} = M(\gamma)$ and $\gamma_i$ is the $i$th integer of the genotype. This is just an example, but it suggests a general approach to handling constraints in "pie" problems.

#### Parameter Optimization Problems

These are problems that involve finding optimal values for a set of parameters of a predetermined mathematical formula: engineering design and numerical regression fall into this class of problems. Evolution strategies (see Section 1.6.1) provide very effective techniques for dealing with this kind of problem.

#### Permutation Problems

These are problems that involve finding a (possibly constrained) permutation of elements: this class comprises some famous problems from operations research like the Traveling Salesman Problem, where an order in which to visit a number of cities is sought for. A wealth of encoding schemes have been proposed for this kind of problem.

There are at least five sensible ways of encoding a permutation of $N$ elements. Suppose $N = 9$; we will identify the nine elements with numbers $1, 2, \ldots, 9$. For instance, we want to represent permutation

$$1-2-4-3-8-5-9-6-7. \tag{1.31}$$

These are some possibilities:

- *adjacency representation:* the genotype is made up of $N$ integers between 1 and $N$ included; a number $j$ in position $i$ indicates that $j$ is the element that comes after element $i$ in the permutation; therefore, the permutation in equation (1.31) would be encoded as $(2, 4, 8, 3, 9, 7, 1, 5, 6)$;

- *ordinal representation:* the genotype is again made up of $N$ integers, but here we imagine starting with all $N$ elements sorted in ascending order; a number $j$ in position $i$ tells us to extract the $j$th elements among the remaining $N-i+1$ and use it as the $i$th element of the permutation; according to this scheme, the permutation in equation (1.31) would be encoded as $(1, 1, 2, 1, 4, 1, 3, 1, 1)$;

- *path representation:* here representation is direct: our sample permutation would be simply encoded as $(1, 2, 4, 3, 8, 5, 9, 6, 7)$;

- *matrix representation:* the genotype is an $N \times N$ matrix of binary digits; a 1 in position $(i, j)$ means that element $i$ comes before element $j$ in the permutation; the permutation in Equation (1.31) would be encoded as

$$\begin{bmatrix} 0 & 1 & 1 & 1 & 1 & 1 & 1 & 1 & 1 \\ 0 & 0 & 1 & 1 & 1 & 1 & 1 & 1 & 1 \\ 0 & 0 & 0 & 0 & 1 & 1 & 1 & 1 & 1 \\ 0 & 0 & 1 & 0 & 1 & 1 & 1 & 1 & 1 \\ 0 & 0 & 0 & 0 & 0 & 1 & 1 & 0 & 1 \\ 0 & 0 & 0 & 0 & 0 & 0 & 0 & 0 & 0 \\ 0 & 0 & 0 & 0 & 0 & 0 & 1 & 0 & 0 \\ 0 & 0 & 0 & 0 & 1 & 1 & 1 & 0 & 1 \\ 0 & 0 & 0 & 0 & 0 & 1 & 1 & 0 & 0 \end{bmatrix};$$

- *sorting representation:* the genotype is an array of $N$ reals or integers, and the permutation is obtained by sorting these numbers in increasing order; a possible representation of our sample permutation would be

$$(-23, -6, 2, 0, 19, 32, 85, 11, 25).$$

Note that according to this encoding scheme, there are many different genotypes that encode for the same permutation.

#### Mapping Problems

Problems that require finding a mapping from a set of inputs to a set of outputs might be termed mapping problems: This class includes, among others, symbolic regression, series prediction, system modeling, and control. Solutions to these problems can be conveniently represented as mathematical formulas or simple programs, for which genetic programming (see Section 1.5) offers a well-established set of specific operators and techniques; alternative representations include neural networks and (fuzzy) rule sets, in which case we fall into the class of problem involving parameter optimization. Most of the rest of the present book will be dedicated to the interaction of these techniques and to their applications.

#### Specialized Genetic Operators

For each encoding scheme like the ones briefly outlined above specific genetic operators have been defined, that operate at the semantic level of the representation. In other words, these specialized genetic operators manipulate elements of a solution in a hopefully sensible way, preserving feasibility of solutions and syntax of the representations.

A detailed treatise of the specialized genetic operators that have been proposed in the literature goes beyond the scope and length of this chapter. What is important for the reader to keep in mind is that the issues of representation and specialized genetic operators are intimately tied and they are one of the frontiers of research on the application of evolutionary algorithms to problems of practical relevance.

### 1.7.3 Handling Constraints

Three techniques have been proposed and used in order to deal with non-trivial constraints:

- the use of penalty functions;
- the use of decoders or repair algorithms;
- the design of appropriate encodings and specialized genetic operators.

Penalty functions are appropriate functions associated with each constraint in the object problem, which measure the degree to which their constraint is violated by a solution. As their name indicates, these functions are combined with the objective function to calculate the fitness of an individual, providing a penalty for each constraint violation. While penalty functions are very general and their application is straightforward, by using them one risks spending most of the time evaluating infeasible solutions, eventually sticking with the first feasible solution found, or finding an infeasible solution that scores better than all feasible solutions.

Decoders are used to translate a genotype into a feasible solution, even when the natural decoding would produce an infeasible solution. Repair algorithms are applied to new individuals generated by mutation or recombination to check whether they violate any constraints and, if that is the case, to "repair" the damage, by transforming their genotype in such a way that it corresponds to a feasible solution. Decoders and repair algorithms are often computationally intensive and have to be tailored to the particular application.

The design of appropriate encodings such that all possible genotypes encode for feasible solutions and/or specialized genetic operators that preserve feasibility is more of an art, requiring some insights into the structure of the object problem. It goes without saying that when such an approach can be taken it results in a much better performance than the other two and a greater elegance.

### 1.7.4 Hybrid Evolutionary Algorithms

Three methods whereby an evolutionary algorithm can be hybridized with available heuristics can be distinguished:

- seed the population with solutions provided by some heuristics;
- use local optimization algorithms as genetic operators;

- encode parameters of a heuristics instead of a solution and then use the heuristics to decode the genotype into its corresponding phenotype.

The first technique is also the simplest. In almost all domains in which an optimization problem arises some sort of preliminary, sub-optimal solution is available, either because an expert has developed it by hand or because some local optimization algorithm has already been used. Care must be taken, however, when seeding evolutionary algorithms with prefabricated solutions, in that the search process might thus be misled into a local optimum and miss better but completely different solutions.

The main motivation behind the second technique is that, as it was observed in Section 1.2.4, as the population starts adapting to the problem, mutation most likely disrupts good solutions rather than further improving them: improving mutations become rarer and rarer. The case for using an available local optimization algorithm as a mutation-type genetic operator is made by the need to improve the algorithm performance by artificially making "good" mutations more frequent. If the locally improved genotype is kept in the new population then the process is commonly called Lamarckian mutation. It is also possible to just keep track of the new and better search point (i.e. phenotype) without incorporating the change into the original genotype.

The third technique is the most sophisticated. It actually implies a total change of perspective – and representation. The search space of parameters of an available heuristics is likely to be much smaller than the space of all solutions to the object problem. The kind of heuristics that typically are available can be characterized as greedy algorithms. These algorithms usually rely on weights assigned to the elements of a solution in order to combine them in the locally best way. If this is the case, those weights can be viewed as parameters of the heuristics: provided that they are assigned in an appropriate way, the greedy algorithm will be able to construct the globally optimal solution. The problem is then shifted to finding an appropriate parameter assignment, rather then, directly, a ready-made solution to the object problem.

### 1.7.5 Co-evolution

In *co-evolutionary algorithms* two or more populations constantly evolve and interact. This is in contrast with the customary evolutionary paradigm where a single population evolves under the selection pressure of a given fixed fitness function that plays the role of the environment. Indeed, in nature the environment of a given population is actually comprised of the physical environment, which normally changes very slowly, and of the other biological populations which are simultaneously adapting. Interactions between (evolving) populations are omnipresent; consider, for example, prey-predator or host-parasite relationships. Under these conditions, it is best to think of evolution as being a co-evolutionary process where changes in a certain species (population) influence the other ones, i.e., the environment is altered. Thus, a kind of "arms race" develops in which evolutionary changes in one species trigger counter-adaptive changes in other species, and *vice versa*.

These observations have only recently been exploited for creating more robust artificial evolutionary algorithms. One advantage of co-evolutionary methods is that one need not necessarily specify a global fitness function, only relative fitness is needed. This can be useful since sometimes providing an adequate fitness function for a given problem can be difficult or even impossible, for example, in complex games or when the suite of test cases is very large.

The methods based on co-evolution can roughly be classified as being either *competitive* or *cooperative*. Hillis presented one of the first successful competitive co-evolutionary approaches to optimization problems [93]. The problem consisted in evolving a sorting network for 16 integers involving a minimum number of exchanges. Hillis used both a classical evolutionary approach and a co-evolutionary one. In the latter there are two populations, the first consisting of sorting networks, the second of sorting problems. These problems are permutations of integers that are to be used as test cases by the sorting networks of the first population; this second population can be viewed as an opportunistic or *parasitic* one. Both populations co-evolve on a two-dimensional grid in parallel, with selection and mating being carried out locally. The fitness for the sorting networks is defined as how well they sort the numbers of the immediate neighbor parasites; the parasites are scored according to their capacity for producing difficult problems for the sorting networks. In this way, the best networks learn to sort increasingly difficult sets of numbers, thus avoiding the need for testing all 16! possible permutations of numbers. The co-evolutionary approach outperformed the standard one for this problem and Hillis was able to evolve a nearly optimal sorting network with 61 exchanges, the best known hand-designed solutions having 60 exchanges.

Potter and De Jong [178, 179] and Husbands [100] have proposed related co-evolutionary models that are also based on species cooperation rather than competition. The methods differ in their implementation but they share the notion of multiple species cooperating so as to attain a common objective. We briefly describe the first approach, referring the reader to the original papers for more details.

In this model, given a problem to be solved, multiple populations are evolved independently by a standard genetic algorithm. Each subpopulation evolves a species of individuals that represent (hopefully) useful components in the solution of the global problem. Species are then combined into full solutions and evaluated on the common global task. Credit is assigned to the species according to how well they collaborate to solve the common problem. In this way, selection pressure favors cooperation rather than competition between species, although within a single species evolution is still competition-based.

Potter and De Jong obtained good results, often better than standard GA-based ones, on simple function optimization problems and on neural network evolutionary design.

Paredis [164] has also studied co-evolutionary dynamics recently, both in the competitive predator-prey setting as well as using a symbiotic approach, which is co-operative in nature, since success on one side improves chances of survival of the other party.

Another highly parallel, local, co-evolutionary algorithm has recently been used by Sipper for evolving non-uniform cellular automata to perform computational tasks. This model belongs to the cooperative class of co-evolutionary algorithms, since the individual units must work in unison to attain a global goal. The methodology, called the *cellular programming method* is described in detail in [210].

## 1.8 A Case Study: Portfolio Optimization

To provide a practical feel of the issues involved with applying evolutionary algorithms to real-world problems, this section examines a financial application, namely multi-period dynamic asset allocation using non-trivial indices of risk.

Multi-period asset allocation is a generalization of one-period portfolio selection. We first review how evolutionary algorithms can be applied to simple, one-period portfolio selection [128] and then move on to the more complex task of multi-period asset allocation.

### 1.8.1 One-Period Portfolio Selection

The central problem of Portfolio Theory concerns selecting the weights of assets in a portfolio that minimize a certain measure of risk for any given level of expected return. According to mainstream Portfolio Theory [136], a meaningful measure of risk is the variance of the distribution of portfolio returns. However, while leading to elegant analytical results and easy practical implementation, this approach fails to capture what an investor perceives as the essence of risk, that is the chance of incurring a loss. In this respect, the economic literature has recently proposed alternative measures of risk; an interesting family of measures, defined as lower partial moments [85], refers to the down-side part of the distribution of returns and has therefore become known under the name of *downside* risk.

A number of difficulties are encountered when trying to apply this new approach even to simple problem instances. First of all, the nature of these risk measures prevents general analytical solutions from being devised and quadratic optimization techniques cannot be applied because the shape of the objective surface is in general non-convex. Moreover, typical size of real-world portfolio selection problems is in the order of several tens or even hudrends of assets. Expected return and risk may be calculated by using historical series made up of hundreds of historical returns, or may be suggested by rather sophisticated financial forecasting techniques.

Even the software packages that apply the conventional mean-variance approach through quadratic optimization suffer limitations as the size of the problem grows beyond a certain threshold. Things become more involved as it is usual practice among fund managers to impose various constraints on their optimal portfolios.

### Downside Risk

In summary, let $\theta$ be a threshold beyond which the investor judges the result of their investment a failure. Adapting the formula for semivariance to this subjective threshold gives the formula of *target semivariance*

$$\mathrm{TS}(\theta) = \sum_{i:W_i<\theta} \Pr(i)(\theta - W_i)^2,$$

where $\Pr(i)$ is the probability of scenario $i$ and $W_i$ is the wealth in that scenario. This measure can be usefully generalized into downside risk, having as parameter a risk aversion coefficient $q \geq 0$:

$$\mathrm{DSR}(\theta; q) = \sum_{i:W_i<\theta} \Pr(i)(\theta - W_i)^q.$$

A higher aversion to risk corresponds to a larger value of $q$.

### Problem Formulation

Portfolio optimization can be viewed as a convex combination parametric programming problem, in that an investor wants to minimize risk while maximizing expected return. This can be expressed as a two-objective optimization problem:

$$\begin{array}{ll} \text{minimize}_{\mathbf{w}} & \mathrm{Risk}(\mathbf{w}) \\ \text{maximize}_{\mathbf{w}} & \mathrm{ER}(\mathbf{w}) \\ \text{subject to} & \\ & \sum_i w_i = 1, \\ & w_i \geq 0. \end{array} \tag{1.32}$$

The set of Pareto-optimal, i.e., non-dominated solutions to this two-objective optimization problem, is called the *efficient frontier*. On the efficient frontier, a larger expected return always corresponds to a greater risk.

The two objectives in Equation 1.32 can be parametrized to yield a convex combination parametric programming problem with objective

$$\underset{\mathbf{w}}{\text{minimize}}\, \lambda\mathrm{Risk}(\mathbf{w}) - (1-\lambda)\mathrm{ER}(\mathbf{w}), \tag{1.33}$$

where $\lambda$ is a trade-off coefficient ranging between 0 and 1. When $\lambda = 0$, the investor disregards risk and only seeks to maximize expected return. When $\lambda = 1$, risk alone is being minimized, whatever the expected return.

Since there is no general way to tell which particular trade off between risk and return is to be considered the best, optimizing a portfolio means finding a whole range of optimal portfolios for all the possible values of the trade-off coefficient; the investors will thus be able to choose the one they believe appropriate for their requirements.

### The Evolutionary Approach

A natural way to achieve the stated objective within an evolutionary setting is to have several distinct populations evolve for a number of trade-off coefficient values. The greater this number, the finer the resolution with which the investor will be able to explore the efficient frontier. Because it is likely that slight variations in the trade-off coefficient do not significantly worsen a good solution, a natural way to sustain the evolutionary process is to allow migration or cross-breeding between individuals belonging to populations corresponding to values of the trade-off coefficient that are close together.

### Encoding

Portfolio selection is a typical example of what we called a *pie* problem (cf. Section 1.7.2) so we can use the straight encoding suggested for this kind of problem: each asset $i$ is encoded as an integer $g_i$ between 0 and 255. The actual weight $w_i$ of asset $i$ is computed as

$$w_i = \frac{g_i}{\sum_j g_j}.$$

### Objective Function and Fitness

The objective function is defined, according to Equation 1.33, as

$$z(\mathbf{w}; \lambda) = (1 - \lambda)\mathrm{ER}(\mathbf{w}) - \lambda \mathrm{DSR}(\mathbf{w}), \tag{1.34}$$

where both return and downside risk are appropriately scaled. By adopting the convention that lower fitness is better, we scale fitness in such a way that the best individual of each subpopulation always has zero fitness. The fitness of individual $\mathbf{g}$, corresponding to portfolio $\mathbf{w}$, of a subpopulation associated with trade-off coefficient $\lambda$ is calculated as

$$f(\mathbf{g}) = \frac{z^{\max}(\lambda) - z(\mathbf{w}; \lambda)}{z^{\max}(\lambda) - z^{\min}(\lambda) + 1}, \tag{1.35}$$

where $z^{\max}(\lambda)$ and $z^{\min}(\lambda)$ stand for the greatest and smallest value of the objective function in the subpopulation.

### Genetic Operators

A variation of uniform crossover, that we may call uniform balanced crossover, was adopted. Let $\gamma$ and $\kappa$ be two parent chromosomes. For each gene $i$ in the offspring, it is decided with equal probability whether it should be inherited from one parent or the other. Suppose that the $i$th gene in $\gamma$, $\gamma_i$, has been chosen to be passed on to a child genotype substring $\zeta$. Then, the value of the $i$th gene in $\zeta$ is

$$\zeta_i = \min\left(255, \gamma_i \frac{\sum_{j=1}^{N} \gamma_j + \sum_{j=1}^{N} \kappa_j}{2\sum_{j=1}^{N} \gamma_j}\right).$$

The main motivation behind this operation is to preserve the *relative* meaning of genes, which depends on their context. Indeed, genes have a meaning with respect to the other genes of the same portfolio, since the corresponding weight is obtained by normalization. Therefore, crossing the two substrings $(1,0,0)$ and $(0,0,10)$ to obtain $(1,0,10)$ would not correctly interpret the fact that the 1 in the first substring means "put everything on the first asset".

A mutation operator that worked quite well on this problem alters a gene either by incrementing or decrementing it by one. This ensures that a chromosome undergoing mutation will not experience abrupt changes.

### Selection and Algorithm

The evolutionary algorithm for one-period portfolio selection is steady-state with elitist selection, with multiple populations connected according to a linear (or stepping-stone) topology, in such a way that adjacent subpopulations have the closest values of $\lambda$.

### Experiments and Results

The evolutionary algorithm described above was packaged, back in 1994, in a decision support system, called DRAGO, on behalf of the mutual fund management branch of a prominent Italian bank. Since its implementation, DRAGO has undergone a comprehensive testing in the field, with portfolios ranging from a few dozens to hundreds of assets, proving quite successful in approximating the efficient frontier for a wide variety of problems.

### A Distributed Implementation

This algorithm naturally lends itself to a distributed implementation (see Chapter 8 for a detailed treatment of parallel and distributed

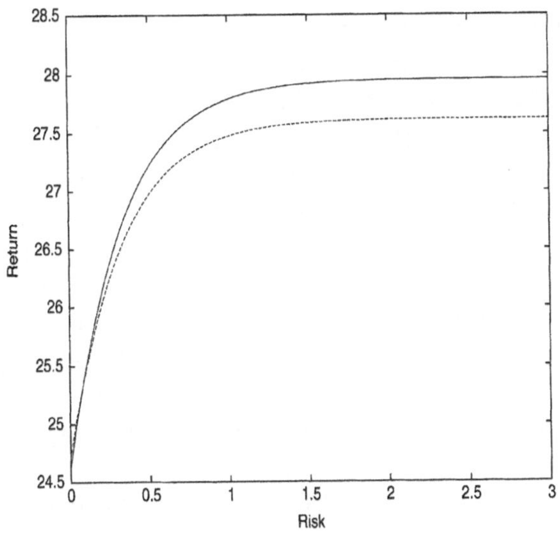


Figure 1.5 Efficient frontier for a 53-asset portfolio. Sequential (dashed line) and distributed (thin line) evolutionary algorithm.

EAs). A distributed evolutionary algorithm was used to solve the portfolio selection problem [129] with very encouraging results. Figure 1.5 shows a fully developed efficient frontier for a portfolio with 53 assets.

It is clearly seen that the distributed version can find portfolio weights that offer a significantly better return for a given risk level over nearly all the risk/return region of interest. We also experimented with very large portfolios (more than 100 assets). In all cases, the distributed GA outperformed the sequential algorithm. In a selection problem with 140 assets the distributed algorithm running on a normally loaded Sun Sparc cluster with 10 machines has been able to develop acceptable solutions in a few hours time, while we had to stop the sequential version after two days, before any reasonable curve had formed.

### 1.8.2 Multi-period Asset Allocation

Portfolio construction can become a very complicated problem, as regulatory constraints, individual investor's requirements, and subjective quality measures are taken into account, together with multiple investment horizons (i.e., multiple periods) and cash-flow planning. This problem may be modeled by building a tree of possible scenarios for the future. An evolutionary algorithm can then be used to optimize an investment plan against the desired criteria and the possible scenarios [15].

Here, instead of considering single assets, we reason at a higher level, by selecting benchmarks for some asset categories (asset classes). These constitute what is called the *opportunity set*.

Traditional portfolio theory models problems where a portfolio is bought at the beginning of the investment and sold once a given time (the investment horizon) has elapsed. Even if, in principle, one could rebalance the portfolio from time to time by using one-period asset allocation techniques, it would not be straightforward to integrate asset decisions and liabilities decisions over time, because one-period portfolio selection neglects information about the likely time paths of the asset classes. The best way to solve the problem involves a dynamic approach, which takes into account the randomness of the time paths of the asset returns.

For this reason, asset allocation is essentially a multi-period optimization problem. A sequence of portfolios is sought which maximizes the probability of achieving in time a given set of financial goals. The portfolio sequences are to be evaluated against a probabilistic representation of the foreseen evolution of a number of asset classes, the opportunity set for investment.

### Problem Formulation

Expected conditions of financial markets, which consist of the possible realizations of returns of asset classes, are described by means of a scenario tree.

The multi-period optimization problem for institutional investors involves maximizing a function measuring the difference between the overall wealth (return) and the *cost* of a failure to achieve a given overall wealth (return), taking into account the cash-flow structure.

Along with objective data and criteria, which are relatively easy to quantify and account for, a viable approach to multi-period asset allocation should consider a number of subjective criteria. These comprise desired features for an investment plan, as seen from the standpoint of the institutional investor and guidelines dictated by regulations or policies of the asset manager. The former are in fact criteria used to personalize the investment plan according to preferences of the customer, even at the expense of its potential return. The latter criteria may consist of specific product requirements or an *ad hoc* time-dependent risk policy.

In order to match subjective advices both from the institutional investors and the asset manager, constraints are implemented by means of penalty functions. In this way, the relative importance of the various constraints can be fine-tuned by the asset manager to ensure a higher degree of customization for solutions.

Overall, the criteria (objectives and constraints) of the multi-period portfolio optimization problem considered can be grouped in five types:

1. Maximize the final wealth;

2. Minimize the risk of not achieving the intermediate and final wealth objectives;

3. Minimize turn-over from one investment period to the next one, in order to keep commissions low;

4. Respect absolute constraints on the minimum and maximum weight for any asset classes or groups thereof;

5. Respect relative constraints on the weight of asset classes or groups thereof.

Optimization is carried out on a portfolio tree matching the scenario tree of returns, considered as a whole. A portfolio tree can be regarded as a very detailed investment plan, providing different responses in terms of allocation, to different scenarios. These responses are hierarchically structured, forming a tree whose root represents the initial allocation of assets. Each path from the root to a leaf represents a possible evolution in time of the initial asset allocation, in response to, or in anticipation of, changing market conditions. The objective function depends on all the branches of the scenario tree.

For each asset class and each asset group (i.e., a group of asset classes) , the minimum and maximum weight is given for portfolios of all periods. In addition, the maximum variation of each asset class with respect to the previous period is given. Beside absolute constraints, relative constraints on asset groups can be defined, such as, for instance, "bond holdings must be more than twice equity holdings".

### Objective Function

The objective function $z$, to be minimized, is given by

$$z = -\mathrm{EW}_{\mathrm{Fin}} + \mathrm{MR} + \mathrm{TC} + \mathrm{WAG} + \mathrm{WRG}. \tag{1.36}$$

The first four terms in Equation 1.36 are the objectives. The first two terms ($EW_{Fin}$, the final wealth, and MR, the mean cost of risk) are the fundamental objectives, i.e. optimization criteria, in that an investor wishes to maximize wealth while trying to minimize risk. The following term (TC, transaction cost) is an optional objective, depending on the valuations of the fund manager. All the other terms are penalties triggered by any violation of a number of constraints. Other constraints can be satisfied "by construction", as explained below, and therefore they need not be associated with a penalty term.

For each constraint, a non-negative weight allows the fund manager to express how much the relevant penalty is to affect the search of an optimal portfolio.

## Encoding

A portfolio tree is encoded as a string of bytes, where each portfolio is encoded by a substring of $N$ bytes (one for each asset class in the opportunity set), starting from the root and visiting all the nodes of the tree breadth-first. While the encoding of the root portfolio is direct, the encoding of all the children portfolios is *differential*. This means that only the changes with respect to the parent portfolio are encoded. The encoding is a generalization and refinement of the straightforward encoding used for the one-period problem.

The decoding of the genotype is performed in such a way that the asset class minimum $W_{it}^{\min}$, maximum $W_{it}^{\max}$, maximum negative variation $D_{it}^{-}$, and maximum positive variation $D_{it}^{+}$ constraints be satisfied for all asset classes $i$ and for all periods $t$.

The decoding of the genotype of a portfolio tree starts by decoding the root portfolio, as follows.

Let $\mathbf{g} = (g_1, \ldots g_N)$ be the genotype substring for the root portfolio; let $\mathbf{w} = (w_1, \ldots w_N)$ and $\bar{\mathbf{w}} = (\bar{w}_1, \ldots \bar{w}_N)$ be respectively the non-normalized weight vector and the corresponding normalized vector for the same portfolio.

It is assumed

$$w_i = W_{i0}^{\min} + \left(W_{i0}^{\max} - W_{i0}^{\min}\right) \frac{g_i}{\sum_{k=1}^{N} g_k}.$$

Therefore, vector $\mathbf{w}$ satisfies the $W^{\min}$ and $W^{\max}$ constraints, but not necessarily $\sum_{i=1}^{N} w_i = 1$. The normalized vector $\bar{\mathbf{w}}$ is obtained from $\mathbf{w}$ as follows:

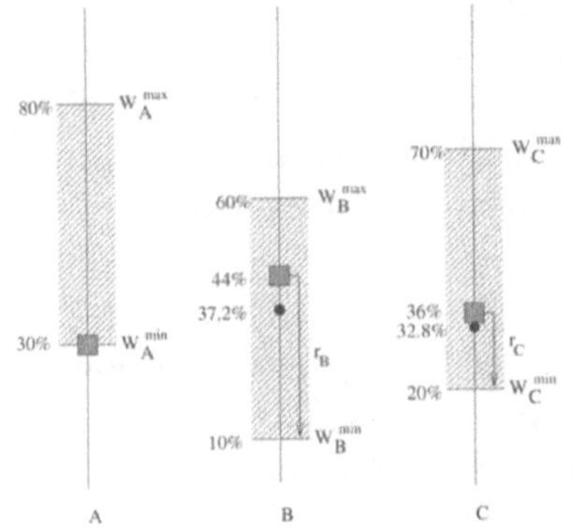


**Figure 1.6** Example of the derivation of a root portfolio for a problem with three asset classes, $A$, $B$ and $C$ and constraints $W_A^{\min} = 30\%$, $W_A^{\max} = 80\%$, $W_B^{\min} = 10\%$, $W_B^{\max} = 60\%$, $W_C^{\min} = 20\%$, and $W_C^{\max} = 70\%$. The feasible weight region for each asset class is indicated by the dashed texture. Assuming that the genotype be the integer vector $\mathbf{g} = (g_A, g_B, g_C) = (0, 15, 7)$, the non-normalized weights $w_A = 30\%$, $w_B = 44\%$, and $w_C = 36\%$, represented as grey squares, are calculated. Since their sum is 110%, $\Delta = -10\%$, while $r_A = 0$, $r_B = -34\%$, and $r_C = -16\%$. The normalized weights, represented as solid bullets, are therefore calculated as $\bar{w}_A = 30\%$, $\bar{w}_B = 37.2\%$, and $\bar{w}_C = 32.8\%$.

Let $\Delta = 1 - \sum_{i=1}^{N} w_i$, and

$$r_i = \begin{cases} W_i^{\max} - w_i & \text{if } \Delta \geq 0, \\ W_i^{\min} - w_i & \text{if } \Delta < 0. \end{cases}$$

The normalized weights are given by

$$\bar{w}_i = w_i + r_i \frac{\Delta}{\sum_{k=1}^{N} r_k}.$$

This procedure is illustrated in Figure 1.6 by means of a three asset class example.

The decoding then continues with the children portfolios, in an analogous way.

## Initialization

The initial population is seeded with random genotypes. This means that every byte is assigned a value at random with uniform probability over $\{0, \ldots, 255\}$.

It is important to notice that, because of the complicated decoding procedure illustrated above, the corresponding portfolio trees will not be uniformly distributed over the space of all feasible portfolio trees. However, the evolutionary algorithm is robust enough not to be negatively affected by this initial bias.

### Genetic Operators

The same crossover used for one-period portfolio selection is still well suited for the multi-period case. The mutation operator is almost the same as in the one-period problem, except for the fact that there is a difference in how random change is carried out between the first $N$ bytes, encoding for the root portfolio, and the rest of the genotype, reflecting the fact that while the former are a direct encoding of a portfolio, the latter encode for the variation in weight from the parent portfolio.

Therefore, the first $N$ bytes are increased or decreased by one with equal probability, while the remaining bytes are completely overwritten by a new random value from $\{0, \ldots, 255\}$. In other words, the first $N$ bytes mutate very gradually, and the others can mutate more widely.

### Fitness and Selection

The fitness of an individual (i.e. of a portfolio tree) is a positive real number $f$ obtained from the objective function $z$ via the transformation

$$f = \begin{cases} \frac{1}{z+1}, & \text{if } z > 0; \\ 1 - z, & \text{if } z \leq 0. \end{cases}$$

Selection is elitist and fitness proportionate, using the roulette-wheel algorithm. Fitness is scaled with the transformation

$$\hat{f} = f - f_{\text{worst}}, \tag{1.37}$$

where $f_{\text{worst}}$ is the fitness of the worst individual in the current population.

Overall, the algorithm is elitist, in the sense that the best individual in the population is always passed on unchanged to the next generation, without undergoing crossover or mutation.

Chapter

2

# Artificial Neural Networks

## 2.1 Introduction

THE FUNCTIONING of the brain has always fascinated people. The human brain, and also the brain of some animals, is indeed capable of astonishing achievements such as remembering, recognizing patterns, and associating, among many others. The way in which these tasks are performed appears to be quite different in nature from standard computation as we know it, that is, in the Turing sense [146]. Indeed, the brain is a massively parallel, highly connected assemblage of an astronomical number of slow processing units that collectively work on these difficult tasks and allow us to function smoothly and effortlessly. These units or cells are called neurons, they are of several different types, and they work in an analog way by propagating electrical currents of chemical origin along connections. The details of how neurons function are very intricate and need not concern us here but the main points are simple and worth some study. The neuron has three main components: the *soma*, the *dendrites*, and the *axon* (Figure 2.1).

The soma is the cell body containing the cell's nucleus. The axon is a single fiber extending from the cell body and then branching out progressively. The dendrites are shorter and they are arranged in a tree-like fashion around the cell body. The axon branches ter-

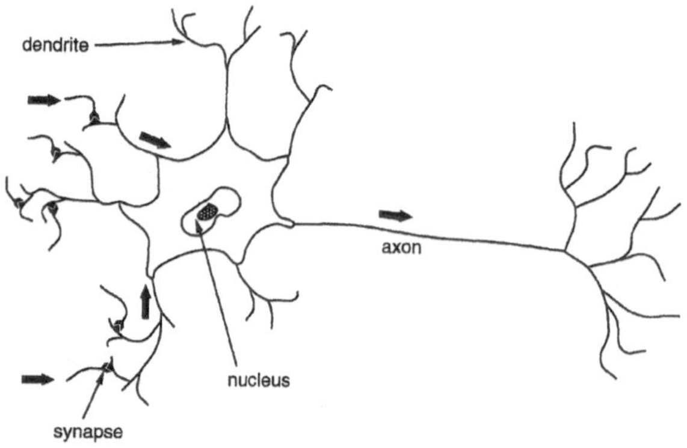


Figure 2.1 A highly stylized form of a biological neuron with its main components highlighted. The arrows show the direction of the electrical signal flow.

minate at the surface of other neurons or on the dendrites. These contact points are called *synapses*. Neurons interact through electrical pulses that propagate along the axons and are transmitted to other neurons via the synaptic connections. If enough incoming pulses arrive at a given neuron in a certain interval of time, the neuron "fires", transmitting a new electrical pulse down its axon. It is this comparatively slow but massively parallel and fault-tolerant signal propagation phenomenon that accounts for the unique properties of the nervous system. The brain is also a low-consumption system, dissipating several orders of magnitude less power than any known digital technology for comparably elementary operations.

The operational properties of the brain arise from several different processes, on different time scales. First of all, the brain has evolved over millions of years and its main structures are coded in the genome. However, given that the human brain has some $10^{11}$ neurons as well as some $10^{15}$ connections, the genetic code cannot specify everything from the start. Therefore, the brain shows "plasticity", that is, it continues to modify its connectivity patterns and to adapt during its whole life, as a function of the external stimuli, although plasticity diminishes with age. It is especially the neuron "fanout", that is, the number of direct connections a neuron has with other neurons, that is impressive, this number being typically between $10^3$ and $10^5$. Just for comparison, this number is of the order of 10 for electronic circuits. No doubt, the highly interconnected nature of the brain plays a major role in the way it functions. The self-organizing and self-adapting capabilities of the brain are also evident in case of impairement of some of the brain area. This also points to another remarkable property of the brain, i.e., its capability to tolerate faults and incomplete information. This fault-tolerance is difficult to obtain in artificial computational systems and the brain is a precious example to follow in this respect.

Artificial neural networks (ANN) have their origin in the attempt to simulate by mathematical means an idealized form of the elementary processing units in the brain and of their interconnections, signal processing, and self-organization capabilities. The emphasis here is on these models seen as computing systems of a different kind. There also exist computational models of real neurons that are widely used in what is called "computational neuroscience". Computational neuroscience tries to build models that approximate the behavior of actual neurons so as to be able to simulate relatively large pieces of the nervous system. This is useful to complement electrophysiological and other in vivo experiments that are difficult and lengthy to perform. But although these models are very interesting and useful, they fall outside the scope of our subject matter. Instead, we intend to present in this chapter the behavior and properties of highly simplified mathematical abstractions of neurons and of their interactions. We will first outline the simplest formal models of neurons, study their properties and limitations as single units, and then investigate the behavior of networks of such simple processing elements. We will see that neural networks solve problems in very different ways from those we are accustomed to in classical computer science insofar as neural networks approximate answers rather than deterministically compute them. In fact, the usual way to solve a given problem on a digital computer is to provide a series of precise instructions to be followed by the machine, i.e., a "program". On the other hand, a neural network may be viewed as an adaptive system that progressively self-organizes in order to approximate the solution. In other words, neural networks free the problem solver from the need to accurately and unambiguously specify the steps towards the solution. This problem-solving philosophy can be either an advantage or a drawback: it all depends on the application and on the objectives.

Most importantly, we will also see that neural networks have the ability to progressively improve their performance on a given task by somehow "learning" how to do the task better, if given some way of evaluating their current performance, a process in which programming is replaced by learning through examples.

Artificial neural networks are at their best for problems in which there is little or incomplete understanding, so that building a faithful mathematical model is difficult or even impossible, but abundant data is available, since they are data-driven. Problems of this kind are very common in pattern classification, non-linear function approximation and system modeling, control, associative memory, and system prediction among others. ANNs of the kind studied here are related to traditional mathematical and statistical models, as we will see in this chapter.

Although this chapter provides a reasonably detailed account of the most common ANN types and gives an adequate background for the rest of the book, it is far from exhaustive. We have concentrated on ANN models that can be used in conjunction with other soft computing techniques, and these combinations will be the subject of subsequent chapters. The subject of ANNs has grown into quite a large field of study and more complete descriptions can be found in a number of books. Among those we can recommend Gurney's book [82] at an elementary level and Hassoun's [86] for a deeper and more complete presentation.

## 2.2 Artificial Neurons

BEFORE studying assemblages of artificial neurons it is sensible to describe the properties of isolated units. The first model of a formal neuron was proposed by McCulloch and Pitts in their landmark 1943 paper [139] and is called a *Threshold Logic Unit* (TLU) or a *Linear Threshold Gate*. Figure 2.2 gives a graphical representation of such a unit with $n$ real-valued inputs $x_i$ each input being associated with a parameter $w_i$. Parameter $w_i$ is also known as a "synaptic weight", or simply "weight", in analogy with biological synapses, the functional contacts between two nerve cells. A TLU performs a weighted sum operation followed by a non-linear thresholding operation, or step function, such that if the value of the sum is greater or equal than the *threshold* $\theta$ then the output $y$ of the unit is 1, otherwise it is 0:

$$y(x) = \begin{cases} 1 & if\ \sum_{i=1}^{n} w_i x_i \geq \theta, \\ 0 & otherwise \end{cases} \tag{2.1}$$

In other words, the neuron will "fire" that is, it will emit an instantaneous "1" signal if the thereshold is exceeded; otherwise, it will do nothing. The weighted sum of equation 2.1, also called the neuron

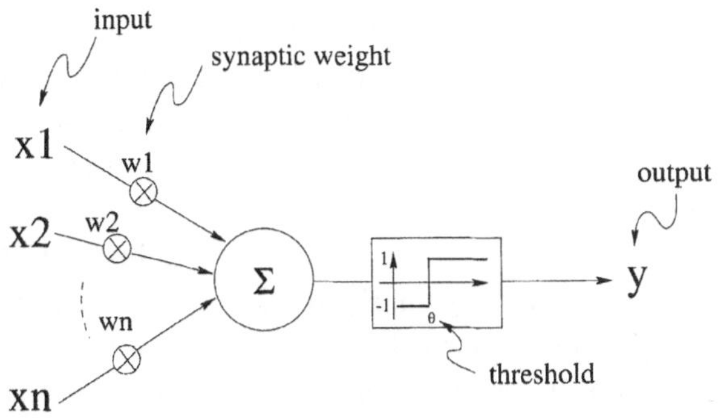


Figure 2.2 The thresholding logic unit of McCulloch and Pitts.

*activation*, can be expressed more concisely as the scalar product $\mathbf{w} \cdot \mathbf{x}$ of the weights vector $\mathbf{w}$ and the input vector $\mathbf{x}$.

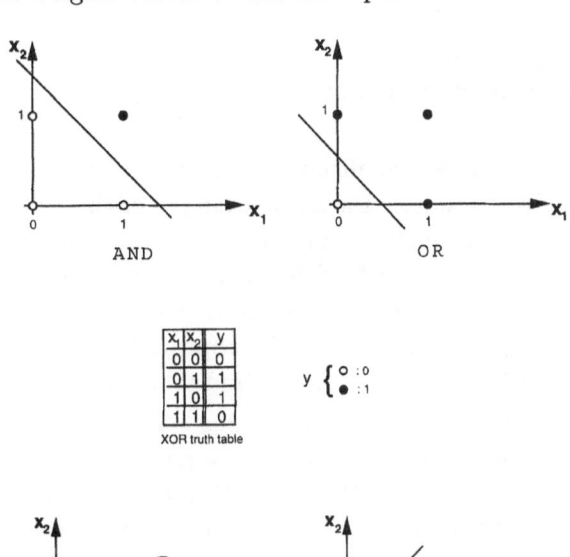

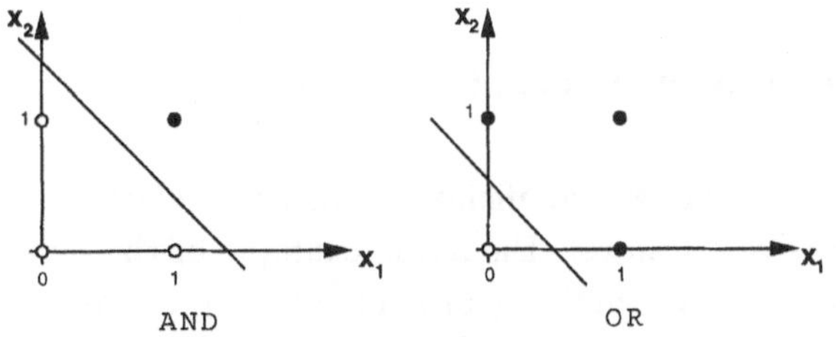


Figure 2.3 Linear separations of input space corresponding to the AND and OR logic functions.

| x1 | x2 | y |
|---|---|---|
| 0 | 0 | 0 |
| 0 | 1 | 1 |
| 1 | 0 | 1 |
| 1 | 1 | 0 |

XOR truth table

y {○ :0, ● :1}

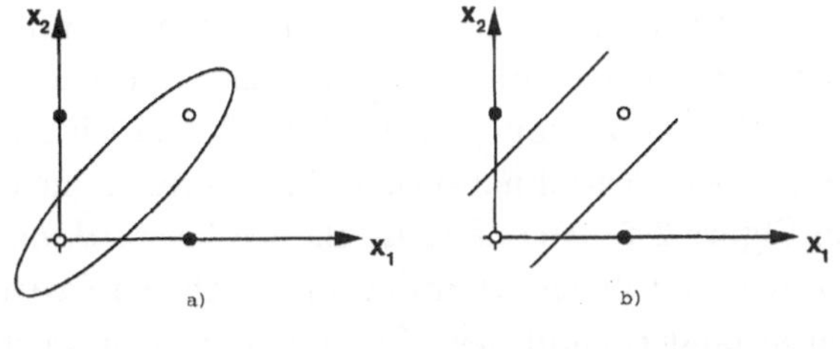


Figure 2.4 XOR implementation with (a) a nonlinear separation of the input space, and (b) two linear separations.

Thus, a TLU performs a mapping $\mathbb{R}^n \to \{0,1\}$ from the reals to Boolean values. If the inputs are binary, that is if the n-component vector $\mathbf{x} \in \{0,1\}^n$ then a TLU becomes a boolean function. It would then be interesting to know the computational power of TLUs, i.e. can a TLU realize all the $2^{2^n}$ possible boolean functions of $n$ inputs? The global answer is that a single TLU can only realize a subset of all those functions (for more details see [86]). The functions that can be attained by a TLU are called *linearly separable* which means that they are able to categorize the inputs in two classes which are separated by a hyperplane in an n-dimensional boolean space. For instance, among the familiar boolean functions with two inputs, AND and OR are linearly separable, whereas the boolean equality and inequality (also called XOR or exclusive OR) are not. This can be seen pictorially in Figure 2.3 and Figure 2.4. In two dimensions a hyperplane becomes a straight line and there is no way of drawing the line in order to separate the two different classes in the XOR case. However, it can also be seen in this last case that a non-linear separator or a couple of straight lines can perform the classification task.

## 2.3 Networks of Artificial Neurons

AS WE HAVE SEEN , artificial neurons in isolation are not very impressive. However, the living example of the nervous system shows that large assemblages of simple cells give raise to astonishingly complex emergent behaviour. Likewise, it is only when many artificial simple units are brought together to form a network that useful new computational capabilities appear. In this section we will study such artificial neural networks and some of their properties.

Although we have just seen that a single TLU is unable to represent all the Boolean functions, it can nevertheless realize the NAND and NOR gates. Since NAND (and NOR) are universal logic gates, it follows that a TLU is also universal and any Boolean function can be realized by a suitable network of TLUs. This result had already been recognized in the classic paper by McCulloch and Pitts [139]. There exist many kinds of neural network architectures, some of which are depicted in Figure 2.7. Among these the *feedforward multilayer networks* have received much attention due to their relative simplicity and computational capabilities. In feedforward networks an input pattern is transformed to an output pattern through a finite series of layers of nodes, some of which may have no connections to the input or output, and there are no feedback signals i.e., signals only travel forward (see Figure 2.7(b)).

For instance, it can be shown that the feedforward two-layer network of Figure 2.5 solves the XOR classification problem by implementing two linear decision boundaries, as depicted in Figure 2.4.

The intermediate node layers are called *hidden* nodes because they are not directly connected to the external world through the inputs and the outputs.

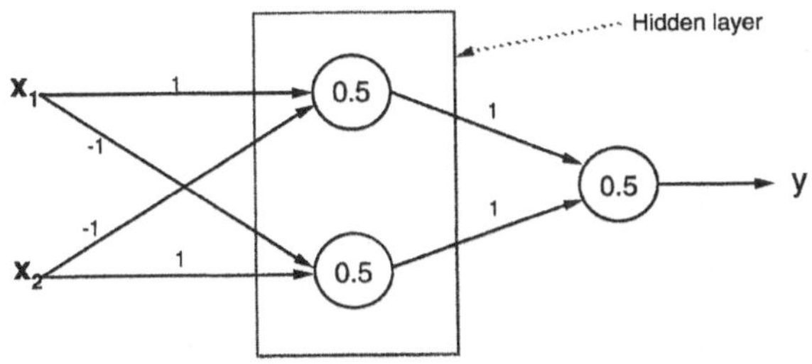


Figure 2.5 Multilayer feedforward network.

Until now we have been mainly talking about Boolean functions as the use of a threshold function limits the output to be 0 or 1. However, it is possible to make the artificial neuron emit a continuous signal. This can be obtained by using a continuous transfer function following the weighted sum instead of the discontinuous step function, as represented in the following equation:

$$y(x) = g\,(\sum_{i=0}^{n} w_i x_i), \tag{2.2}$$

where $g(w_i x_i)$ is called the *activation function*. A convenient function form for $g$ is the so-called *sigmoid* which is graphically represented in Figure 2.6 and whose analytic form is:

$$y(x) = \frac{1}{1 + e^{-(bx-c)}} \tag{2.3}$$

Other commonly used continuous transfer functions are linear, hyperbolic tangent, and gaussian. The sigmoid is suitable because it is differentiable and it saturates i.e. it tends asymptotically to 0 and 1 at the extremes. By varying the parameter of the exponential one can control the steepness of the curve and in the limit it reduces to a TLU. Note that still other functions can be used provided that they are continuous, monotonically increasing, and take values between 0 and 1.

Networks of neurons with real-valued inputs and a sigmoid transfer function can be used to approximate mathematical functions. In fact, it is possible to think of a network of such units as implementing a mathematical function of its inputs. This allows the parameterization of an unknown real-valued function which is very useful in situations where the exact form of the functional relation (if any)

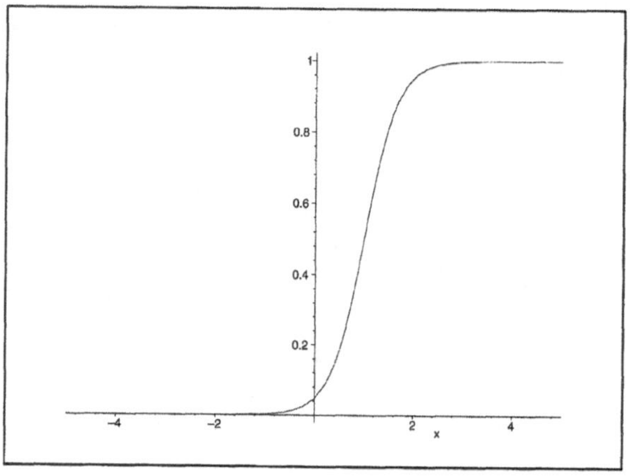


Figure 2.6 A typical sigmoid activation function.

is unknown, such as time-series prediction or system identification. More generally, let us suppose that we want to approximate any continuous real-valued function $F(x_1, x_2, \ldots, x_p)$. A general result (see for instance [86] and references therein) says that such a family of functions can be approximated to any desired accuracy by a feed-forward network with at least one single hidden layer. That is, there exist real constants $\alpha_i, w_{ij}$ and a monotone-increasing continuous function $g$ (e.g. the sigmoid function) such that:

$$f(x_1, x_2, \ldots, x_p) = \sum_{k=0}^{p} \alpha_k \, g(\sum_{j=0}^{m} w_{ij} v_j) \tag{2.4}$$

and

$$\mid f(x_1, x_2, \ldots, x_p) - F(x_1, x_2, \ldots, x_p) \mid < \epsilon \tag{2.5}$$

for any $\epsilon > 0$, where $m$ and $p$ are the number of units in the input and hidden layers respectively. Being only an existence result, the theorem leaves unspecified the optimum number of hidden layers and the number of units in the hidden layers in any given case. A related result is that single-hidden-layer nets with sigmoidal activation units and a single linear output node are universal classifiers, i.e. they can correctly assign any given input pattern to one of a finite number of distinct classes. In practice, networks used for classification usually employ several output units, each of which represents a unique class. This setting is more natural, as it avoids the use of a large number of hidden units.

Thus, ANNs possess interesting computational capabilities. But the remaining problem is that of designing the networks in such a way that a given computational task can be realized as efficiently and economically as possible. How should we go about designing a neural network? How many units and layers should we use? How to choose the connections and their weights? There are no easy answers to these questions and no standard design recipes exist for designing neural networks for a given problem. In general, neural network researchers prefer to let the network design itself, so to speak, rather than imposing a pre-existing architecture. This is indeed a very peculiar way of tackling a computational problem and it will be dealt with in the following section.

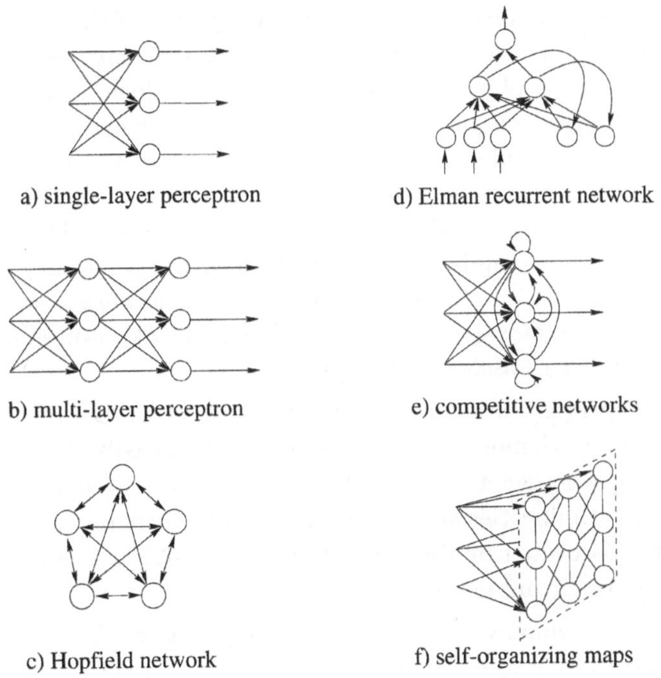


Figure 2.7 Typical neural network topologies.

## 2.4 Neural Learning

WE SAW that one of the distictive features of artificial neural networks is the absence of a pre-defined set of instructions to follow. Instead, neural networks represent a different conceptual approach to computation that makes use of statistical concepts such as progressively improving input-output relationships by a process in many ways analogous to function fitting and extrapolation. In this way, the network adapts itself to perform a given task. In the neural network literature such a process goes under the name of "learning" or "training". Perhaps, terms such as "learning" are misleading since they may suggest that a phenomenon similar in complexity to human or animal learning is taking place. Indeed, nothing of the sort is actually occurring when a network is trained to perform a given task. The process is entirely of a formal nature and can be described by well-defined mathematical algorithms. Nevertheless, network and machine learning is such a widespread concept that the term will be used here in this restricted sense.

In its most common meaning, learning by a neural network implies an adaptive procedure in which the weights of the network are incrementally modified so as to improve a prespecified performance criterion i.e., an objective function over time. Such a procedure is called a *learnig rule* or *learning algorithm* and the adaptation may take place in a *supervised* or an *unsupervised* way. In supervised learning the net is presented with a set of known input/ouput pattern pairs: this set of values is called the *training set*. The learning process consists in updating the weights at each training step so that, for a given input, an error measure between the network's output and the known target value is reduced. This is also known as learning with a teacher or associative learning.

In unsupervised learning there is still an input/output relationship that the network must learn to reproduce but no feedback is provided indicating whether a given association is correct or not. In other words, the similarities among patterns and features of a training set must be discovered by the network itself by *clustering* similar cases in a *self-organizing* manner.

A third commonly used form of training a network makes use of the concept of *reinforcement learning*. In this case the "teacher" signal for some training input/output pair is not some measure of the difference between the given output and the expected value, as in supervised learning, but rather an evaluation of the result as a "wrong" or "right" direction.

In the following sections some of the most important supervised and unsupervised learning algorithms will be briefly described. Since the subject is a large one, we cannot hope for complete coverage: for an in-depth treatment of the subject, the reader can consult for instance reference [86].

## 2.5 Supervised Learning

SUPERVISED learning algorithms are based on error correction rules, that is, an error value is generated from the actual response of the network and the desired response. Following that, the weights are modified such that the error is gradually reduced.

We start by describing some simple single-neuron training rules and then go on to training algorithms for networks of units.

### 2.5.1 Perceptron Learning Algorithm

A classical and very simple example of supervised learning is Rosenblatt's *perceptron learning algorithm* [190]. For our purpose here, a perceptron is a binary unit similar to the linear threshold gate of Figure 2.2. The outline of the learning algorithm is given in Figure 2.8.

1. Initialize weights and threshold randomly,
2. Present an input vector to the neuron,
3. Evaluate the output of the neuron,
4. Evaluate the error of the neuron and update the weights according to:
   $w_i^{t+1} = w_i^t + \eta(d - y)x_i$,
   where, d is the desired output, y is the actual output of the neuron, and $\eta$ $(0 < \eta < 1)$ is a parameter called the *step size.*
5. Go to step 2 for a certain number of iterations, or until the error is less than a prespecified value.

Figure 2.8 Perceptron learning algorithm.

Historically, the algorithm is called the perceptron learning rule because it was first used by Rosenblatt with a variant of the TLU called a perceptron. In this algorithm, the positive parameter $\eta$ is called the *learning rate* or step size: it dictates the size of the weight change and thus it controls how fast the learning takes place.

The rule has a straightforward geometric interpretation in terms of vector algebra in $\mathbb{R}^n$ but we do not have space here to go into the details. To give a feeling for how the algorithm works, let us consider a decision problem in which the inputs must be partitioned into two classes. In the perceptron context, learning is equivalent to changing the vector of weights in such a way that a decision surface moves until a hyperplane is found that correctly separates all input/output pairs into two distinct classes. The interested reader can consult [82] for a more formal explanation.

Rosenblatt proved that if the two classes of input patterns are linearly separable, then the perceptron algorithm will eventually converge after a finite number of iterations. This is known as the *perceptron convergence theorem*. An important property of the perceptron is that whatever it can compute, it can learn with the perceptron learning algorithm. However, if a given problem is not linearly separable, a perceptron does not find a low-error solution, even if it exists.

The perceptron learning rule is of historical interest but it fails on nonlineraly separable problems. We have seen in Section 2.3 that such nonlinearly separable problems can be solved in principle by multi-layer nets of LTUs. However, it turns out that multi-layer nets cannot be trained effectively by the perceptron learning algorithm. For training purposes, it has been found that instead of using straight weight adjustment, it is more practical to use gradient-based descent on an appropriate differentiable criterion function. For instance, we could use as an objective function to be minimized the sum of the differences between the actual neuron output and the target training output values. This amounts to finding the minimum of the sum of errors over the training set as a function of the free weight parameters. Normally, the square of the sum of the errors is used in order to work with positive quantities. Because of the discontinuous nature of the output step function, the gradient, and thus the derivatives of the criterion functions are also discontinuous. This is easy to fix by just considering the neuron activation i.e., the weighted sum of inputs instead of the output value. As we will see, this form of training is suitable for multi-neuron, multi-layer networks, allowing those networks to solve arbitrary classification problems (see also Section 2.3).

### 2.5.2 LMS Rule and Delta Rule

The LMS (*Least Mean Square*) algorithm by B. Widrow and M. Hoff [245] was developed to provide "learning" to a McCulloch-Pitts-like element called ADALINE for ADAptive LINear Element. Such nodes are identical to TLUs except that the output signals are $\{-1, +1\}$ instead of $\{0, 1\}$. It uses a hard-limiting function with a bias weight $w_0$ controlling the threshold level of such a function. The learning rule tries to reduce the mean square error (MSE), averaged over the training set, using the *gradient descent* method. Here the criterion function to be minimized is:

$$E(\mathbf{w}) = 1/2 \sum_{i=1}^{m} (d_i - y_i)^2 \tag{2.6}$$

That is, the sum of squared errors over $m$ training pairs, where $\mathbf{w}$ is the vector of weights. Taking the derivatives gives us the iterative form of the rule:

$$\mathbf{w}^{t+1} = \mathbf{w}^t + \mu\,(d_k^t - y_k^t)\,\mathbf{x}_k^t, \tag{2.7}$$

where $\mu$ is a parameter that controls stability and rate of convergence, $\mathbf{x}$ and $\mathbf{w}$ are the input and weight vectors respectively, $y$ is the output value, $d$ is the expected value and $t$ and $t+1$ are the current and next step in the iteration respectively.

In the LMS algorithm the weights are updated using an estimate of the steepest descent of the mean square error function $E(\mathbf{x})$ in weight space. $E$ is a quadratic function of the weights and is therefore convex and has a unique (global) minimum. If the chosen positive constant $\mu$ is sufficientely small, the gradient descent search will asymptotically converge toward the solution regardless of the initial weight values. Widrow extended his ADALINE units to multiple Adaline networks and provided one of the first trainable layered networks.

The so-called *Delta* rule is similar to the LMS rule: it also works by minimizing an error residual but now the activation function of the neuron unit is a continuous, differentiable, non-linear curve such as a sigmoid. Gradient methods for minimization can be used with the Delta rule and the output value can appear in the quadratic criterion function since the transfer function is differentiable. The convergence speed of the Delta algorithm depends on the slope of the first derivative of the transfer function, which can be rather flat at the extremes. In these regions progress is very slow but convergence can be improved in various ways (see for instance [86]).

The importance of the Delta rule is due to the fact that it extends naturally to the training of multilayer networks by a method called *error backpropagation.* Backpropagation was worked out early in the 1970s by Paul J. Werbos. The next section gives an account of this important technique.

### 2.5.3 Backpropagation Algorithm

The absence of a practical method for training multilayer networks nearly stopped work on ANNs for several years. The invention of the backpropagation method, attributed to Paul J. Werbos [239], is one of the main reasons for the renewed interest in artificial neural networks toward the end of the 1970s.

The algorithm is by far the most popular method for performing supervised learning of feedforward networks composed of continuous activation function units such as those described in the previous paragraph on the Delta rule in order to be able to use derivatives for calculating the gradient. It has been used on countless applications of ANNs to many different kinds of problems. Error backpropagation is essentially a search procedure that attempts to minimize a whole network error function such as the sum $E$ of the squared error of the network output over an ensemble of training input/output pairs:

$$E = 1/2 \sum_{j=1}^{m} (d_j - y_j)^2, \tag{2.8}$$

where $d_j$ is the desired $j$th output and $y_j$ is the actual $j$th output of the network

The name of the algorithm is due to the fact that the weight modifications dictated by the learning rule propagates "backwards" from the output layer to the input layer. In fact, the algorithm can be intuitively visualized as a series of forward and backward waves of activity. In the forward phase the network produces its output for given input/output pairs: this leads to the calculation of the global error. In the backwards phase, the weights of the output nodes and then those of the hidden nodes back to the input, are modified layer by layer according to the learning rule in order to reduce the error. The details of the algorithm are mathematically simple but require some space to be presented in an orderly way. We give here only an outline of the basic procedure (Figure 2.9). An exhaustive description can be found in [86]. An excellent pictorial explanation is contained in [188].

1. Initialize weights randomly,
2. Present an input vector pattern to the network,
3. Evaluate the outputs of the network by propagating signals forwards,
4. For all output neurons calculate $\delta_j = (y_j - d_j)$, where $d_j$ is the desired output of neuron $j$ and $y_j$ is its current output: $y_j = g(\sum_i w_{ij}x_i) = (1 + e^{-\sum_i w_{ij}x_i})^{-1}$, assuming a sigmoid activation function,
5. For all other neurons (from last hidden layer to first), compute $\delta_j = \sum_k w_{jk}g'(x)\delta_k$, where $\delta_k$ is the $\delta_j$ of the succeeding layer, and $g'(x) = y_k(1 - y_k)$,
6. Update the weights according to: $w_{ij}(t+1) = w_{ij}(t) - \eta y_i y_j(1 - y_j)\delta_j$, where, $\eta$ is a parameter called the *learning rate*.
7. Go to step 2 for a certain number of iterations, or until the error is less than a prespecified value.

Figure 2.9
Backpropagation learning algorithm.

### Remarks on the Backpropagation Algorithm

Today there exist many public domain or commercial packages implementing some form of feedforward ANNs with backpropagation as the learning algorithm. There is nothing wrong with this "canned" software as many of them are of high quality and the method is quite standard and well-known. Indeed, in view of the number of applications that use supervised neural learning, it would not make much sense to reinvent the algorithm each time. Nevertheless, there are a number of subtle points in training ANNs that make life less pleasant and that should at least be briefly mentioned.

### Local and Global Minima

First of all, consider that backpropagation essentially does a gradient descent search through the multidimensional space of possible weights trying to reduce the error $E$ between the training output values and the network outputs. Till now, we have not explicitly mentioned the fact that the search hypersurface will almost surely be multi-modal i.e., it may present a number of minima. In minimizing $E(\mathbf{w})$ for such a function we would like to strive for the *global* minimum, that is the value $\mathbf{w}^*$ of the weight vector for which $E(\mathbf{w}^*) < E(\mathbf{w})$ for any $\mathbf{w}$ in the search domain. Gradient descent is a minimization algorithm that will find the closest local minimum with respect to the starting point of the search, and if the search landscape is complex enough, this minimum is very unlikely to be the global one. What does this mean in terms of the learning process? Two things can be said in general about this state of affairs: first, the problem of the search getting stuck in a local minimum is not as important in practice as one might think. Network weights determined in this way have proved to be good enough in many applications. Secondly, there exist other optimization methods that can help the search to escape local minima. For example, evolutionary algorithms (see Chapter 1), which can be seen as biased stochastic optimizers, can be used to train neural networks of various kinds. This is a first example of the synergy between different soft computing methodologies which is one of the leading themes of our book. Other global optimization methods, such as stochastic gradient search or simulated annealing can also be used. Another problem with the standard backpropagation algorithm is that it can be very slow. But researchers have found a number of ways to speed-up and enhance backpropagation learning (see for instance [86]).

### Learning Topologies

Another issue which we have largely ignored until now is the question of network topology. That is, backpropagation or similar learning algorithms are customarily used simply for adjusting the weights in an otherwise fixed feedforward network structure. But it is clear that the interconnection of the units and their number plays an important role. In fact, lack of knowledge in determining the appropriate topology for a problem, including the number of layers, the number of neurons per layer, and the interconnection scheme, often results in slow learning speed and poor performance. For example, if a network is too small, it does not accurately approximate the desired input to output mapping. If a network is too large it requires longer training times and may not perform well on unseen data. A variety of methods have been proposed to dynamically grow or prune the number of neurons and interconnections in order to improve the performance of the network. For example, the *cascade-correlation* algorithm starts with a network with no hidden units and keeps adding units one at a time, choosing the weight such that the residual error is minimized [60]. One can also go the other way around, starting with a complex network and shrinking it by pruning units away when it is found that certain connections have a small influence on the network error. Some of the methods in this class alter the topology while at the same time using backpropagation of the output errors, some do not; but we do not have space here to go into the details, for which the reader can consult [10] for example.

We note here that one major application of evolutionary algorithms to ANNs has been evolving neural network topologies. This topic, together with other interactions between ANNs and EAs will be the theme of Chapter 4.

### Generalization and Overfitting

*Generalization* means the ability of a learning system to correctly map new inputs that were not previously used in the training phase. If we let the network adapt too well to the training data by giving it too many degrees of freedom, then the residual error will be very small but the network is likely to fail to correctly map new, previously unseen input data of the same class i.e., it will have poor generalization capabilities (see Figure 2.10). This phenomenon is known in the statistical and ANN literature as *overfitting* the data and should obviously be avoided as far as possible. The problem is that the net adapts too much to the noise and fine behaviour present in the data while ignoring the underlying general trends.

What can be done to improve the generalization ability of a neural network? One well-known technique comes from regression analysis in statistics and consists in dividing the training data into two sets:

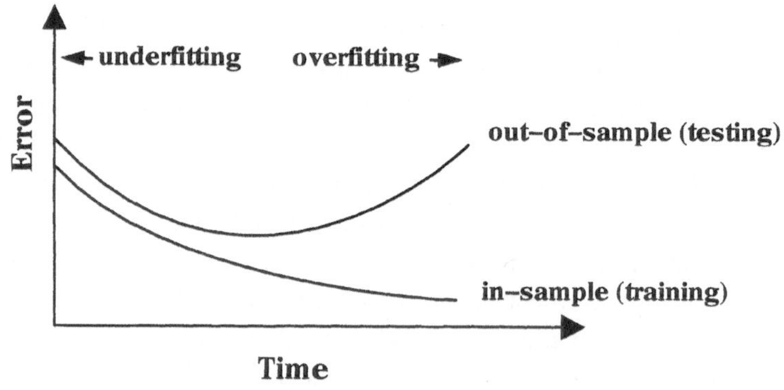


Figure 2.10 Illustration of overfitting. Too much training may lead to small residual error but with bad generalization performance.

a training set proper and a *validation* set. The training set is used for learning in the usual way but from time to time the validation set is used to test out-of-sample performance. This cross-validation procedure continues until the error on the validation set starts to increase, for this means that the generalization performance attained is good. Ideally, the best model is the one at the minimum of the validation set error curve. Cross validation is a useful empirical technique that allows us to distinguish between signal and noise in the absence of further information, and its proper use is the subject of much discussion.

The appropriate number of hidden units and the number of training data are also related and play a role in the quality of the resulting network. There are no rigorous results here but a rule of thumb says that the number of hidden units needed will increase as the number of training data increases for a given level of performance. The techniques outlined in the previous section may help in finding a nearly optimal or satisfactory net topology for a given task.

### 2.5.4 Applications of Supervised Learning Networks

Supervised neural networks, especially feedforward nets trained by backpropagation, have been used in a wide range of applications including signal and image processing, speech recognition, system identification, medical diagnosis, financial prediction, detection of events in high-energy physics, etc.

It would be inappropriate to treat representative applications in detail in this introductory chapter. However, a brief description of a couple of typical areas in which ANNs have been successful should help the reader by giving a flavor of the work that has been achieved.

### An Illustrative Example: Character Recognition

Fast and reliable character recognition, especially of hand-written characters is extremely important in practice. Think for instance of pen-based palm organizers, of automatic address reading and routing of postal mail and of check signatures, just to mention a few. There have been several attempts to solve this problem by using feedforward ANNs. It is very easy to find plenty of training data in this field but the task is a very difficult one even for human recognizers, not to speak of computer programs. Given the difficulties of the problem, no classification system is expected to achieve a success rate of 100 percent. Hundred percent success is not really needed for the system to be useful, but too low a recognition rate, say under 95%, will result in a system that is not reliable enough to be usable in practical applications. An example of this was the Apple Newton, a palmtop computer that employed handwriting input and recognition, but in which error rates were sufficiently high as to seriously limit the user acceptance of the device.

The main ideas in using feedforward nets for character recognition are simple enough: once one has obtained the hand-written data, the characters are scanned and digitized as a gray-scale pattern represented as a grid of pixels of a certain size. Two examples of such a grid are schematically depicted in Figure 2.11. Note, though, that actual grids are likely to have many more pixels in them.

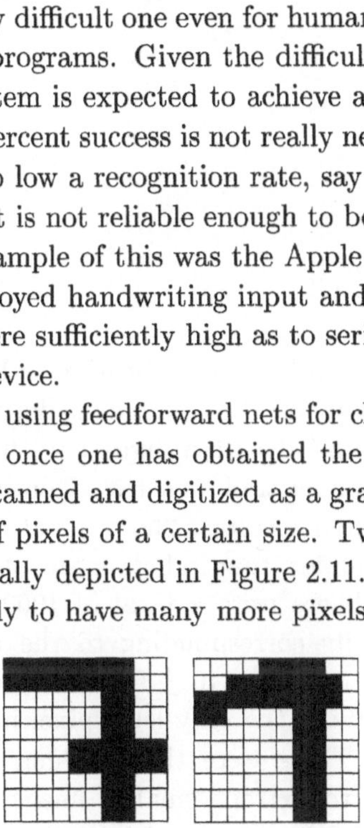

Figure 2.11 An array of pixels representation of two digits. Left hand side: the idealized digit 7. Right hand side: a more likely example of a handwritten or noisy 7.

For characters to be interpreted on personal digital assistants, the recording is made online using a special pad which is able to sense the pen's pressure and velocity. The data are then divided into a learning and a test set. The learning set is presented to the network one image at a time and the error in classification is progressively reduced by using backpropagation. When an acceptable error level has been attained and misclassifications are rare, the net is used to classify the yet unseen examples. The structure of an idealized ANN architecture for decimal digit recognition is shown in Figure 2.12. The inputs are the pixels representing a single character and the outputs are the classes $0 - 9$ into which the given character is to be classified.

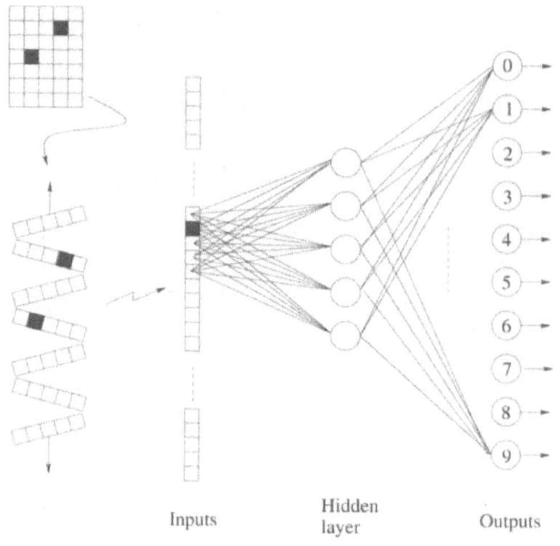


Figure 2.12 A schematic network architecture for character classification.

For example, in a well-known application to handwritten numerical ZIP codes appearing on U.S. mail, a feedforward net was trained to learn to recognize handwritten digits [49]. The ANN learned on several thousand samples and was tested on about two thousand unseen cases with good results. The original network architecture is actually more complex than what one could imagine from our short description. There were a total of 1000 units of which ten were the output units corresponding to the classification of the digits into one of ten groups $(0 - 9)$. The network input was a $16 \times 16$ array containing the gray-level pixel image of a particular handwritten digit. There were three hidden layers of which the first two were able to extract smaller image features, while the third was a standard fully connected layer. The net was trained by an accelerated version of backpropagation. As a refinement of the above work, LeCun *et al.* [49] did weight pruning using statistical methods in order to make a nearly optimal selection of weights with the goal of increasing the generalization capability of the network. This resulted in better overall performance of the system. For more recent discussions of neural and fuzzy approaches to handwriting recognition see [73, 45].

### A Second Example: Financial Markets Forecasting

The dynamics of financial markets is poorly understood. Although market data are generally and publicly available, nobody really knows how market prices will evolve in spite of claims to the contrary. The fact is that financial markets are very complex evolving systems of many interacting agents that have their own beliefs, expectations, and trading strategies. Apart from textbooks describing extremely simple and idealized cases, there is absolutely no market model available. Again, when there are no models but empirical data abound, ANNs, among other techniques, could be employed to approximate the unknown relationships underlying the observed values.

Being able to predict market evolution, at least the trend during a specific time window, is obviously a definite advantage for traders. Some people believe that the future behaviour of a given market is somehow "implied" by the current and past market data and that the careful study of these data patterns will reveal future trends to some extent. Others think that only the fundamental economic indicators such as a nation's GNP (Gross National Product), unemployment rate, and the like have an influence on markets. Yet others think that market indicators such as the yen/dollar ratio are the really important data to take into account. Finally, there are those for whom the market is not predictable at all, not even in principle. For them, prices essentially follow a random walk and any information is immediately absorbed into the current prices and cannot be made use of.

Today, there are technical reasons to believe that markets are not entirely random, at least on some time scales, although the question is far from settled [58]. If one accepts this view that there are opportunities for predicting future prices, what are the possible approaches? One could use econometric models such auto-regressive moving averages of prices plus random shocks to try to forecast future behaviour. But these models are normally linear ones, while non-linear and even chaotic behaviour has been reported for financial markets [58]. Thus, it would seem that feedforward multilayer ANNs might be useful, since they are able to map any well-behaved function of the inputs thanks to their universal approximation capabilities (see Section 2.3).

Feedforward networks trained by backpropagation can be employed to achieve non-linear forecasting of time series. The general idea exploits a result of Takens [86] and it goes as follows: let

$$p(t), p(t-1), p(t-2) \ldots p(t-m) \tag{2.9}$$

be the current and past $m+1$ values of an observation such as a price or the sampled value of a signal. Using these values as an input pattern to a neural network and the observed value $p(t+1)$ at time step $t+1$ as a target for many values of $t$, we can train a feedforward network to make one-step ahead prediction of the unknown future value of the time series. Prediction $n$ steps ahead can also be done but the quality of the prediction will degrade as $n$ increases. Once the net has learned to make the prediction on the training values, it can be used out-of-sample for predicting future values of the series. What the network learns implicitly is of course an approximation of the unknown functional relationship between the next value and the $m$ past samples of the variable. Reported results using this prediction method are at least as good as several other methods for non-linear forecasting [238]. This univariate, pure time-lagged model works well for physical systems showing well-defined deterministic chaos. In the less precise world of economical prediction one could also exploit the multivariate approach that consists in taking into account related time series to the one for which a predictor is sought and also some important market indicators as input values. Such an approach has been implemented with good effect in [6, 7], where a feedforward neural network is used to find a mapping $f$ of the form:

$$x(t) = f(I_1, I_2, \ldots, I_n, x(t - \Delta_i)) \tag{2.10}$$

where $x(t-\Delta_i)$ are lagged values of the variable to be predicted and the $I_j$ are financial indicators, possibly lagged, that might have an influence on $x$ according to economic theory. Among the many possible indicators, the most influential ones are chosen by using statistical sensitivity analysis methods. For instance, in a prediction study of the Swiss stock market index (SPI) [6], it was found that the most relevant indicators are the monthly differences of long interest rates in Germany and Japan and the money exchange rate between USA and Japan. These indicators, together with the past samples of the SPI time series are fed as inputs to a two-layer network. The output is the one-step ahead prediction of the SPI value.

Fairly good results have been obtained with ANNs in the financial markets, in general better than those obtained with most other methodologies (see for instance [58] and references therein). However, one should be aware that there are no "magic" solutions in this difficult field, only promising techniques. The phenomena are not well understood and the ANN approach, even when it works, has a limited temporal validity due to the evolutionary character of the markets. Besides, the well-known black-box character of ANNs calls for using them together with other methodologies having more explanatory power.

## 2.6 Unsupervised Learning

IN PREVIOUS sections we discussed networks of units in which the nets were trained to perform an unknown input/ouput mapping by presenting the network examples of input/output pairs. In unsupervised learning there is no feedback from the environment as to the correcteness of the mapping; in other words, there is no "teacher". Instead, the network must be able to discover by itself any categories, patterns, or features possibly present in the data. Networks that are able to infer pattern relationships without being supervised are also called *self-organizing*. Among the unsupervised learning rules, we will briefly discuss here Hebbian learning for single units and networks, competitive learning, and Kohonen's self-organizing feature maps.

### 2.6.1 Hebbian Learning

The rule that goes under his name, was proposed by Donald O. Hebb in his seminal work *"The Organization of Behavior"* [87]. Hebb's rule makes the weight strength proportional to the product of the firing rates of the two interconnected neurons. That is, when two connected neurons fire at the same time and repeatedly, the synapse's strength is increased. This biologically-motivated rule can be expressed for a single unit as :

where $x_i^t$ and $y_j^t$ are the output values of neurons $i$ and $j$ at time $t$, $w_{ij}^t$ is the current interconnection weight between neuron $i$ and $j$, and $\rho$ is a parameter called the *learning rate*. $w_{ij}^{t+1}$ is the future value of the synaptic weight being updated during learning. Hebb's rule, which has inspired a large number of learning algorithms has the

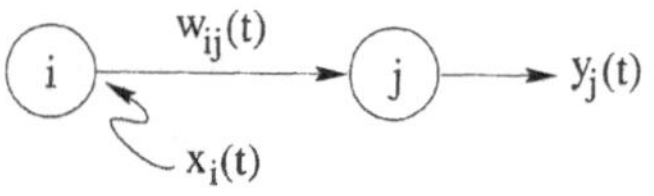


$$w_{ij}^{t+1} = w_{ij}^{t} + \rho\, y_j^t\, x_i^t, \tag{2.11}$$

important characteristic that it is local. The original Hebb rule is divergent, progressively driving the weight to an infinite magnitude. Several modified Hebbian rules that do not have this undesirable property have been devised [86].

Hebbian learning can be applied to networks of interacting units. One of the most studied approaches is called Principal Component Analysis (PCA). PCA is a standard statistical technique whose goal it is to extract $m$ normalized orthogonal vectors in the input space that account for as much of the data's variance as possible. By projecting the n-dimensional input vectors onto the $m$ ($m < n$) orthogonal directions, we can achieve dimensionality reduction with minimum loss in information content. Such a transformation is achieved by rotating the coordinate system with standard linear algebra operations. E. Oja [160] demonstrated that an artificial neural network is able to compute in parallel, and on-line, the PCA transform. The PCA network is a layer of parallel linear artificial neurons. Oja's algorithm is a modification of Hebb's rule in which a weight-decay term proportional to $y^2$ has been added. In Oja's network PCA takes place as a consequence of unsupervised neural learning with a Hebb-like learning rule in which weight updating for a network is done as follows:

$$w_{ij}^{t+1} = w_{ij}^{t} + \rho\,(x_j^t - \sum_{k=1}^{m} w_{kj}^t y_k^t)\, y_i^t, \tag{2.12}$$

where $\rho$ is a learning constant and the other symbols have the usual meaning. There is not room to further develop these ideas here but the interested reader can consult [86] for instance.

### 2.6.2 Competitive Learning

Competitive learning is an unsupervised learning procedure in which the neurons of a network learn to recognize clusters of similar input vectors. The network detects regularities and correlations among the input vectors and adapts the future response of the units to similar inputs. In competitive networks output units compete among themselves for activation. The simplest competitive learning network consists of a single layer of output neurons to which all inputs are connected. All the units are presented with a given input vector $\mathbf{x}$ but only one output neuron is activated at any given time: the so-called *winner* neuron. The winner unit $i$ is the one with the largest activation value (see Section 2.2):

$$\mathbf{w}_i \cdot \mathbf{x} \geq \mathbf{w}_k \cdot \mathbf{x}, \ \ \forall k \neq i \tag{2.13}$$

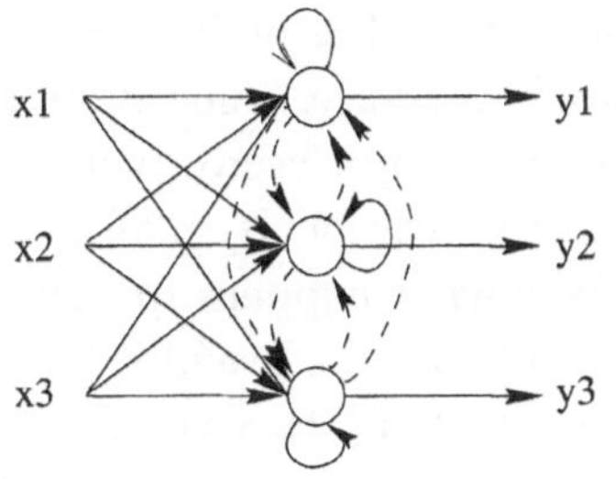


Figure 2.13 A single layer competitive learning network. The solid lines indicate excitatory connections whereas the dashed lines indicate inhibitory connections.

If the input vectors are normalized ($\| \mathbf{x}_k \| = 1$, for all $k = 1 \dots n$) then the unit $i$ with the smallest activation:

$$\| \mathbf{w}_i - \mathbf{x} \| \leq \| \mathbf{w}_k - \mathbf{x} \|, \ \ \forall k \neq i \tag{2.14}$$

that is, the unit with normalized weight closest to the input vector, becomes the winner.

The winner-take-all operation is implemented by connecting the outputs to the other neurons by means of so-called *lateral inhibitions*, and also by means of *self-excitatory* connections as depicted in Figure 2.13. Reference [86] and the references cited therein give details of how inhibition and self-excitation is effected.

In practice, there is no need for the above net to be implemented explicitly: the winning neuron can be found by simple search of the maximum activation. The neuron having the maximum activation updates its weight while the weights of the other neurons remain unchanged according to the iterative application of the following rule:

$$\Delta w_{ij} = \begin{cases} \rho(x_j - w_{ij}) & \text{if } i \text{ is the winning unit} \\ 0 & \text{otherwise} \end{cases} \quad (2.15)$$

The training process can be seen as a progressive tilt of the weight vector of the winning unit towards the direction of the current input vector. Normalized input vectors (see above) can be depicted as points on the surface of a unit sphere (or circle). At the beginning the unit's weight vectors and the input data vectors are not aligned, if one assumes that both the initial weights as well as the input values are drawn from a random distribution. With time, the competitive learning rule tends to associate certain units with neighboring input vectors, also called data *clusters*. This phenomenon is geometrically represented in Figure 2.14. When a stable solution is found then the weight vector for each cluster represents in some sense the "center of gravity" of that cluster, a kind of typical vector for this class of data. Since the number of data clusters is not known in advance, some units in the network, that happen to be far from any input vector, will turn out to be useless, in the sense that their activation is small or zero for the data of the training set. This is what happens to vector $\mathbf{w}_4$ in Figure 2.14. These "spare" units could still be useful if the distribution of input vectors changes in time, making the system more robust. Otherwise, there are several ways to turn these units into "useful" units during the learning process (see [86]).

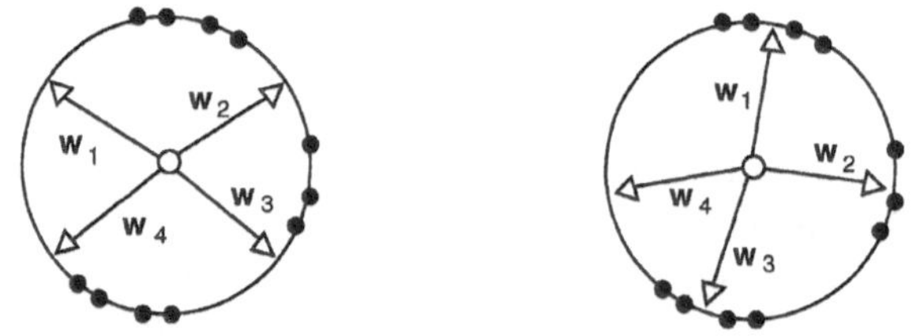


Figure 2.14 A schematic representation of unsupervised competitive learning. The points on the circle surface represent training data. On the left hand side of the figure initial unit weights are represented as random vectors. After training, the weight vectors orient themselves toward the data clusters (right hand side figure).

### Vector Quantization

An important use of competitive learning networks is *vector quantization* which has applications in data compression. In this scheme the input space is divided into disjoints regions in such a way that any input vector $\mathbf{x}$ falling into one of the regions is represented by a single label characterizing that class. The class label encoding can then be used later instead of the vector itself, leading to efficient compression of data for storage and transmission purposes. The classes are defined by a number of prototype vectors and the class to which a given input belongs is calculated by taking the nearest prototype vector using a Euclidean distance metric.

There also exist supervised versions of vector quantization, such as Kohonen's *learning vector quantization* [86].

## 2.6.3 Self-Organizing Features Maps

Biological neural networks often show an architecture that depends on their function. For example, in the visual region of the mammalian cortex the receptive zones are arranged in such a way that neighboring light photons will stimulate areas of the cortex that are also physically close to each other. This is called a *topographic map* and is quite a common arrangement of neurons in regions of the cortex that are responsible for processing sensory information such as visual and audio stimuli. Several observations point to the fact that such topographic maps are not entirely genetically determined. Rather, they seem to be created during the individual development by a sort of unsupervised learning process.

Teuvo Kohonen invented an ANN model directly inspired by the existence of such topographic maps in the brain [114]. Kohonen's self-organizing map (SOM) is an unsupervised learning network in which the neurons have a spatial arrangement, i.e., the neurons are typically organized in a line or a plane (see Figure 2.15). This kind of network is able to encode proximity features among the data. In other words, nodes that are neighbors in the network encode patterns that are adjacent in some sense in pattern space. Indeed, a self-organizing map has the property of *topology preservation*, that is, nearby input patterns should activate nearby output neurons on the map. A network that performs such a mapping is also called a *feature map*. In self-organized maps the weights of locally interacting units are modified in response to input vectors according to a certain rule until a global ordering emerges.

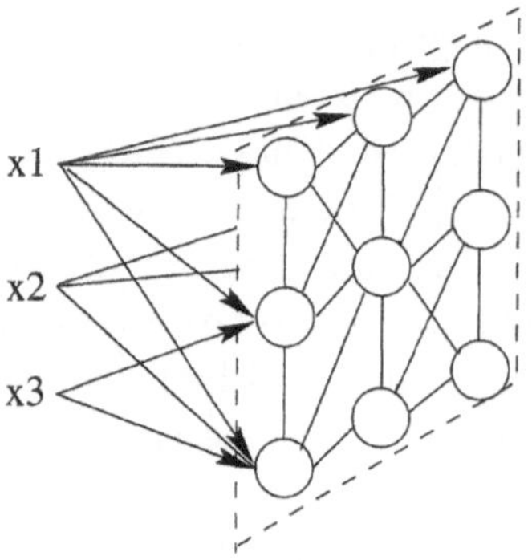


Figure 2.15 Self-Organizing Map.

An early model of self-organization was developed by Willshaw and von der Malsburg in the mid 1970s [235]. They used excitatory lateral connections with neighboring neurons, and inhibitory connections with distant neurons, instead of using a strict winner-take-all mechanism, as shown in Figure 2.16. The function that defines the form of such lateral connection weights is known as the *Mexican hat.*

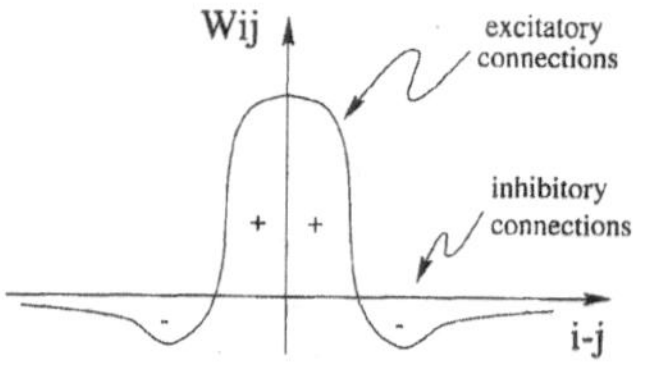


Figure 2.16 Mexican hat form of the lateral connection weights.

Teuvo Kohonen [114] described a self-organizing map learning algorithm that approximates the effect of a Mexican hat form of lateral connection weights by taking into account neighboring units for each output neuron. Neighborhoods introduce the notion of topology in neural learning and make the neurons' position with respect to each other a significant aspect of the net. The neighborhood can be square, rectangular, circular, etc. During training, all weight vectors associated with the winner neuron and its neighbors are updated.

SOM can be seen as an extension to the simple competitive network of the previous section. Given an input vector **x**, the winner unit is determined in the same way as in Equation 2.14. Weight adaptation is then performed according to a learning rule very similar to the competitive rule of Equation 2.15:

$$\Delta w_{ij} = \begin{cases} \rho N(i)(x_j - w_{ij}) & \text{for all units in the neighborhood} \\ 0 & \text{otherwise.} \end{cases} \tag{2.16}$$

The difference with the simple competitive rule is the presence of a "neighborhood" function $N(i)$ of unit $i$. This function is usually symmetric and monotonically decreasing with the distance. At the beginnings of the learning process $N$ defines a large neighborhood and all the units in the neighborhood are updated for any input vector $\mathbf{x}$. As time goes by and learning progresses, the neighborhood is shrunk down while the learning rate $\rho$ is also being iteratively decreased in order to achieve convergence. With respect to competitive training, learning now takes place over an extended neighborhood and it is this feature that allows the emergence of a topographic map. Eventually, a topographic map will be obtained in which nodes not only represent clusters of data but they are also arranged in the map such that neighboring units encode clusters that are "close" to each other in some sense in the input pattern space.

An example of this process is a two-dimensional self-organizing map of 30 neurons used to classify a certain number of random two-dimensional input vectors between -1 and 1. The two-dimensional map is a five by six neurons grid. Since each unit has only two inputs, it is possible to visualize the representation of this net in weight space, in which a point corresponds to each node's weight vector. The progressive organization of the weights according to the topology of the input space is depicted in Figure 2.17. At the beginning the weight vectors are randomly placed. However, after about 300 cycles one can see that the map has begun to organize itself and, as time passes and the neighborhood shrinks, the map is more and more evenly distributed across the input space. Finally, after 1000 cycles, the net has become ordered with the correct topology. Since the input vectors covered the square uniformly, the final map is correspondingly an almost regular grid in this domain. In this sense it can be said that the map has been able to "learn" the topology of the input space.

Feature maps have been applied in many areas including speech recognition, motor control, robot sensing, finance, data compression, and combinatorial optimization.

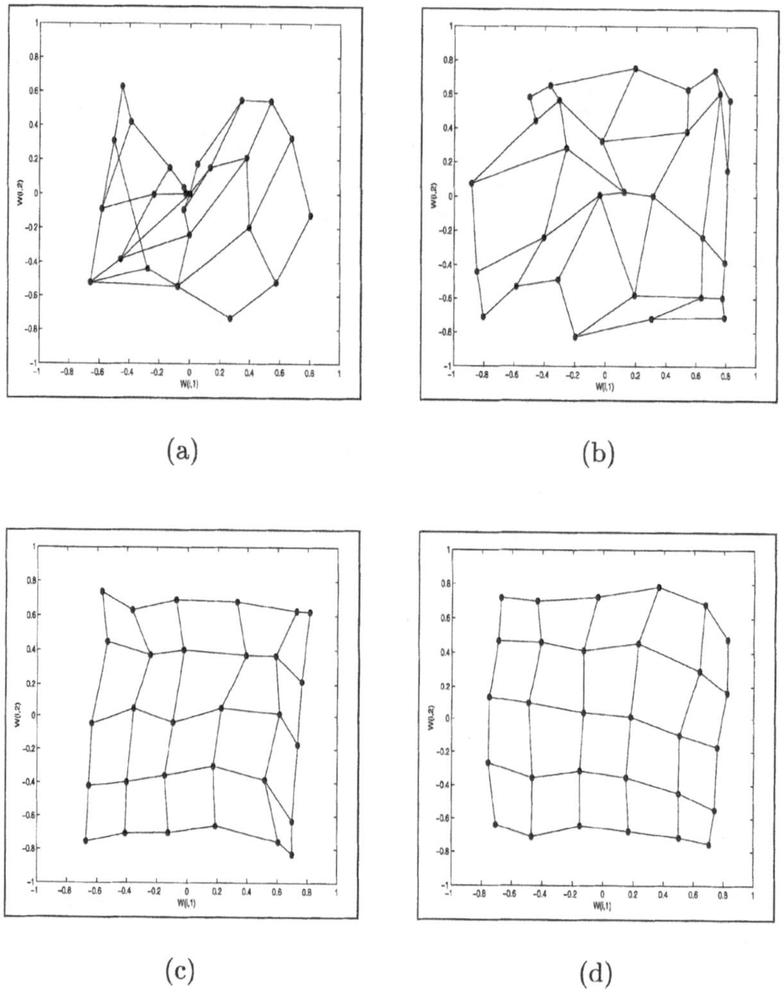

Figure 2.17 A 6 × 5 planar array of units mapping input vectors uniformly distributed in the interval [-1,1] . (a) After 100 iterations. (b) After 300 iterations. (c) After 500 iterations. (d) Final configuration after 1000 iterations. These images have been obtained with the Matlab software.

## 2.7 Fault Tolerance

SOFT-COMPUTING methodologies enjoy a certain amount of robustness in the face of errors, and this aspect is one of their strong points when applied to noisy and imprecise environments. Although ANNs are inspired by the nervous system, we have seen that they are but a pale imitation of actual biological neural networks. However, they retain some of the outstanding features of the latter such as collective computation abilities and resilience with respect to noise and errors. The fault-tolerant aspects of neural networks are all the more appreciable if one consider how faults may affect classical computing devices. Many computational procedures are strictly deterministic: given certain input values, they will always output the same results, using the same state transitions. This fact is comforting but what happens if errors appear somewhere? Change a single bit in that lovely fast sorting program and chances are that it won't work anymore. The fact is that engineered systems, as well as classical algorithms are built in such a way that errors always have serious or fatal consequences, unless much redundancy is built into the system. Thus, the standard engineering design process produces systems that are optimized in some sense but that are also very fragile.

In contrast, the human brain, in spite of its enormous complexity is much more reliable and less prone to catastrophic failure. Individual neurons are constantly dying without being replaced but the brain manages to compensate for this loss during lifetime, unless the damage is a serious one. Natural systems have not been designed in the customary engineering way; rather, they are the result of the evolutionary process (see Chapter 1). As such, they have been able to survive in all kinds of hostile environments by selecting and reproducing the structures that were the fittest at a given moment. At the same time, new structures were being created continuously by mutation and recombination and tried in new environments. These largely random processes have been remarkably successful, although they have taken millions of years. The human brain is as much the product of evolution as it is of learning. In fact, the detailed "wiring" of the brain is too complex to be determined by the genetic code alone. Together with nerve cell maturation and morphogenesis, learning shapes and fine-tunes the cell connections, allowing the brain to become the marvelous machine that allows us to function properly in a continuously changing and complex environment. We will see in subsequent chapters that ANNs too can be designed by similar processes of evolution and learning, but these processes will obviously be artificial and usually simulated by a computer program. Artificial neural networks too can withstand a certain amount of errors and units may fail without necessarily driving the whole system to a grinding halt. This is also called *graceful degradation* and is obviously a very desirable property for machines as well as for humans. Most of the ANN systems that have been described in this chapter are not very sensitive to small amounts of noise or the loss of a unit here and there. Weight noise and unit misfiring has even been used purposefully during ANN training in order to enhance the fault-tolerance properties of the nets [48]. If the artificial neural system is a physical machine and not just a simulation program, this feature is even more useful, since failures are possible and very real in this case.

## 2.8 Artificial Neural Nets and Statistics

IN many fields such as social science, finance, and economy there often exist a lot of data but explanatory theories and models are either lacking or they are not realistic enough as to guide prediction and decision. In data-rich but model-poor environments, people normally use statistical inference techniques to build system understanding and to estimate models. Parametric and non-parametric linear and non-linear regression models, such as autoregressive moving average (ARMA) models are well-known examples of statistical inference methods. Non-linear, nonparametric methods in particular make few *a priori* assumptions about the underlying system and, as a consequence, have a considerable number of degrees of freedom and explore a large space of functions to fit the observations.

It is clear that there are close links between neural network models and these statistical analysis techniques. Table 2.1 gives a minimal glossary of terms describing similar concepts in the two fields.

Indeed, from the formal statistical point of view, ANN models such as supervised feedforward nets can be seen as non-linear regresssion models [44], while other models are similar or identical to clustering, discriminant analysis techniques, or to stochastic approximation theory. Thus, the question is: are these ANN models really new and

| Neural Networks | Statistics |
|---|---|
| learning | model estimation |
| supervised learning | non-linear regression |
| unsupervised learning | cluster analysis |
| weights | parameters |
| inputs | independent variables |
| outputs | dependent variables |

Table 2.1 Artificial neural networks and statistics: glossary of corresponding terms.

what do they have to offer with respect to established statistical methods?

The answer (see [44] and the comments by other authors therein) is not entirely clear. As noted, it appears that, indeed, ANN formalisms can be reduced for the most part to some known statistical model. But for sure, ANNs are "sexier" and more attractive than equivalent statistical techniques for many non-specialists. Also, statisticians have concentrated for the most part on linear models and on comparatively small typical numbers of parameters. Neural networks provide us with tractable multivariate non-linear models that are easy to implement and that enjoy suggestive pictorial interpretations, avoiding the somewhat "dry" aspects of an equivalent statistical approach. There are other advantages as well: they are simple to implement, they enjoy a great simplicity of representation, and can be easily tuned to particular problems. Besides, they can be implemented in hardware with the obvious performance gains due to the direct execution and their intrinsic parallelism. Another advantage is that ANNs can be used as "modules" in *hybrid* systems. Statistics, on the other hand, may give ANNs researchers a firm footing for evaluating and analysing new ANN techniques and for relieving somewhat the lack of explanatory power of ANNs. Neural networks and statistics should not be seen as competing methods. Combinations of ANN and purely statistical methodologies are of great interest and are being actively pursued [44].

Chapter 3

# Fuzzy Systems

## 3.1 Introduction

FUZZY SET THEORY was initiated by Lotfi Zadeh in the mid-sixties. In 1965 Zadeh, then chair of the Electrical Engineering Department at the University of California at Berkeley, published a paper called "Fuzzy Sets" [257].

In the decade after Zadeh's seminal paper, many theoretical developments in fuzzy logic took place in the United States, Europe, and Japan. Then, since the mid-seventies, Japanese researchers have taken a leading role in advancing the practical application of the theory.

### 3.1.1 Uncertainty and Imprecision

Fuzzy set theory provides a mathematical framework for representing and treating uncertainty in the sense of vagueness, imprecision, lack of information, and partial truth.

Very often, we lack complete information in solving real world problems. This can be due to several causes. First of all, human expertise is of a qualitative type, hard to translate into exact numbers and formulas. Our understanding of any process is largely based on imprecise, "approximate" reasoning. However, that imprecision does not prevent human beings from performing tasks very well that no computer (the *precise* machine by excellence) has been able to attempt up to now, such as driving cars and all kinds of vehicles, governing complex engines, and even supervising other people. Furthermore, the main vehicle of human expertise is natural language, which is in its own right ambiguous and vague, while at the same time being the most powerful form of communication ever discovered.

Requiring precision in models and products translates into requiring many resources, both in terms of money and time. For situations other than very simple ones, expense is more than linearly proportional to precision. This necessarily means that in any case a fair trade-off needs to be found between cost and precision, resources employed and accuracy obtained. One of the main motivations of fuzzy set theory and logic is to make this balance of cost and precision explicit and to try and exploit the tolerance for imprecision to reduce cost.

### 3.1.2 Fuzziness and Probability

Discussions about the relation between fuzzy logic and probability are still numerous and sometimes rather controversial. They mainly originate from the fact that probability theory has been until recently the only formal framework for reasoning about uncertainty. Therefore, the most frequent objection to fuzzy set theory raised by probabilists may sound like "we already have a mature and successful theory of uncertainty, so why bother to invent another"?

Actually, fuzziness and probability both deal with uncertainty, but as a matter of fact they deal with two different types of uncertainty. Imagine being lost in a desert, with no water left. At a certain point, you find two packs of 100 bottles of water each. However, there is a trick. On one pack, there is a label saying: "Attention: one of these bottles contains a deadly poison which tastes and looks just like water!". On the other pack, the label reads: "Attention: the water contained in these bottles is polluted: one deadly dose of poison is diluted in the whole pack". You have to choose one bottle to drink: from which pack do you take it? The first pack is a typical situation to which chance applies: there is a $\frac{1}{100}$ probability of drinking from the wrong bottle and dying. The second pack is an example of fuzziness: no matter what bottle you drink from, the water will be slightly poisonous and you will end up having some gut problems, but you will definitely and certainly survive. In the first case the water is either pure or deadly, but there is uncertainty as to which of the two cases will happen; in the second case the water can be anything from pure to deadly, with all possible gradations. Its degree of purity is fuzzy.

The issue of relations between fuzzy logic and probability is treated from a logical standpoint by [83]. In recent years, a consensus has been growing within the soft computing community that fuzzy set theory and probability theory should be viewed as complementary and that the latter needs an infusion of the former to enhance its ability to deal with real world problems [262].

## 3.2 Fuzzy Sets

THIS SECTION provides an introduction to fuzzy set theory, which is at the base of fuzzy logic and fuzzy systems. Classical elementary set theory will be quickly reviewed, then its shortcomings will be pointed out with the help of a few simple and intuitive examples, making the case for the introduction of the notion of degree of membership to a set and, therefore, for a more general definition of set, the fuzzy set.

Several good books are available for reviewing this basic material (see for example [56, 113, 264]), hence we will not enter into too much detail here.

### 3.2.1 Classical Sets

It is a remarkable fact that the formal definition of such an intuitive concept as that of a *set* can be dangerously prone to paradox. The most famous of them is Russel's Paradox: naïve set theory falls into a contradiction while attempting to define the set $\Omega$ of all sets that are not members of themselves. If $\Omega$ were not a member of itself then it would be a member of itself and *vice versa*. Yet, despite paradoxes like this, so far the only workable definition of a set is the intuitive one.

Intuitively, a set is any collection of elements. Examples of sets can be: all the straight lines on the plane, the letters of the alphabet, the people taller than 180 cm, and so on. Sets with a finite number of elements can be described by explicitly listing all their elements between curly braces with no particular order: for example, the set of all positive divisors of 6 is $\{1, 2, 3, 6\}$, which is exactly the same as $\{1, 3, 6, 2\}$.

More formally, $x \in A$ stands for $x$ is an element of set $A$ and $x \notin A$ means that $x$ is *not* an element of set $A$. Since a set is completely determined by its elements, two sets $A$ and $B$ are equal if and only if they have the same elements, i.e. $A = B$ is equivalent to saying that, for all $x$, $x \in A$ if and only if $x \in B$. By this definition, $A \subset B$ and $B \subset C$ implies $A \subset C$.

Given two sets $A$ and $B$, their *intersection* $A \cap B$ is the set of all the elements shared by $A$ and $B$:

$$A \cap B = \{x : x \in A \text{ and } x \in B\}.$$

Their *union* $A \cup B$ is the set of all the elements that are in $A$ or in $B$ or both:

$$A \cup B = \{x : x \in A \text{ or } x \in B\}.$$

The set containing all meaningful or relevant elements in a given context (for instance a problem or an argument) is called the *universe of discourse* and plays a special role both in conventional set theory and in fuzzy set theory, enough to reserve the letter $U$ for it. The *complement* of a set $A$, denoted $\bar{A}$, is the set of all elements in the universe of discourse that do not belong to $A$:

$$\bar{A} = \{x : x \notin A\}.$$

Finally, the *difference* of $A$ with respect to $B$, denoted $A \setminus B$, is the set of all the elements of $A$ that do not belong to $B$ too:

$$A \setminus B = \{x : x \in A \text{ and } x \notin B\};$$

the reader will notice that the difference can be defined using intersection and complement: $A \setminus B = A \cap \bar{B}$. Dual to the universe of discourse is the empty set, that is the set that contains no elements, denoted by the symbol $\emptyset$. Two sets $A$ and $B$ are said to be *disjoint* if their intersection is empty, that is if, $A \cap B = \emptyset$.

The operations defined above satisfy a number of identities, or properties. Some of them are important because they define rules for the mathematical manipulation of sets, like the following:

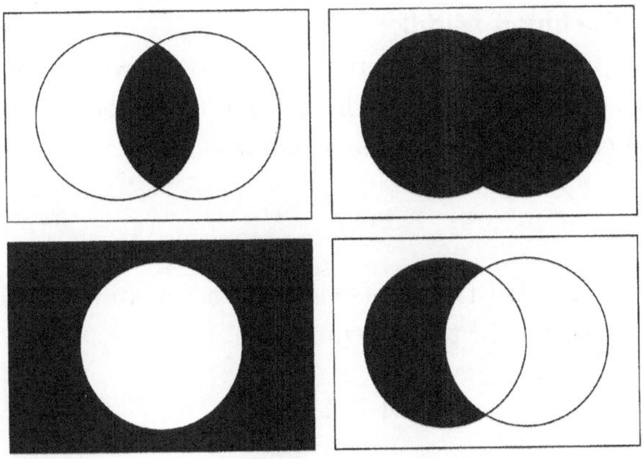

Figure 3.1 Diagram representation of the intersection, union, complement and difference operations on classical sets.

- Commutativity:

$$A \cup B = B \cup A;$$
$$A \cap B = B \cap A.$$

- Associativity:

$$A \cup (B \cup C) = (A \cup B) \cup C;$$
$$A \cap (B \cap C) = (A \cap B) \cap C.$$

- Distributivity:

$$A \cup (B \cap C) = (A \cup B) \cap (A \cup C);$$
$$A \cap (B \cup C) = (A \cap B) \cup (A \cap C).$$

- Idempotency:

$$A \cup A = A;$$
$$A \cap A = A.$$

- Identity:

$$A \cup \emptyset = A;$$
$$A \cap U = A;$$
$$A \cap \emptyset = \emptyset;$$
$$A \cup U = U.$$

- Transitivity:

$$\text{if } A \subseteq B \subseteq C, \text{ then } A \subseteq C.$$

- Involution:

$$\overline{\overline{A}} = A.$$

### Law of the Excluded Middle

Two special properties of set operations are known as the *excluded middle laws*. The first deals with the union of a set and its complement and is called the *law of the excluded middle*:

$$A \cup \bar{A} = U;$$

the second deals with the intersection of a set and its complement and is called the *law of the contradiction*:

$$A \cap \bar{A} = \emptyset.$$

These two laws were one of the foundations of Aristotelian logic and were expressed by the famous latin phrase *tertium non datur*, "a third possibility is ruled out", meaning that an element must either belong to $A$ or to $\bar{A}$.

### De Morgan's Laws

De Morgan's laws are important because they are useful in proving tautologies and contradictions in logic, as well as in many other set operations and proofs. They are described by the following equations:

$$\begin{aligned} \overline{A \cap B} &= \bar{A} \cup \bar{B}; \\ \overline{A \cup B} &= \bar{A} \cap \bar{B}. \end{aligned}$$

As a matter of fact, these laws establish a duality relation between the intersection and union operations.

### The Characteristic Function

The *characteristic function* of a set $A$, is a mapping

$$\chi_A : U \to \{0, 1\}, \tag{3.1}$$

such that, for all $x$,

$$\chi_A(x) = \begin{cases} 1, & \text{if } x \in A; \\ 0 & \text{otherwise.} \end{cases} \tag{3.2}$$

All operations on sets have an alternative but equivalent definition in terms of characteristic functions: for all $x \in U$,

- intersection: $\chi_{A \cap B}(x) = \min\{\chi_A, \chi_B\}$;

- union: $\chi_{A\cup B}(x) = \max\{\chi_A, \chi_B\}$;

- complement: $\chi_{\bar{A}}(x) = 1 - \chi_A$;

and so on. Equality and containment relations can be defined using characteristic functions as well:

$$A = B \Leftrightarrow \text{for all } x \in U, \quad \chi_A(x) = \chi_B(x);$$
$$A \subseteq B \Leftrightarrow \text{for all } x \in U, \quad \chi_A(x) \leq \chi_B(x).$$

### 3.2.2 Shortcomings of Classical Sets

Classical sets have served science very well for centuries. However, there are many real-world situations in which the simplistic vision of reality they are based upon shows its limits. A few examples should make this point clearer.

Imagine having defined the set of full glasses. We are given an empty glass: it certainly does not belong to the set of full glasses. We then begin adding small drops of wine. At what point does the glass begin to belong to the set of full glasses?

Along the same line of reasoning is Zeno's paradox. Imagine having a heap of sand. If you remove a grain of sand you still have a heap of sand. However, if you keep on removing one grain at a time for long enough, eventually you will be left with none, and the heap will no longer be there. When does the heap cross the boundary from heap to non-heap?

Two people, John and Mary go to the same bank asking for a loan. Their salaries are almost the same: John makes \$24,998 a year, whereas Mary makes \$25,000. Unfortunately for John, however, the bank's policy is not to give loans to people making less than \$25,000. That is, the bank partitions its customers in two sets: customers who make \$25,000 a year or more and customers who do not. The result is that Mary gets her loan, while John does not. Is the bank doing a good deal denying John his loan? Or, viewed from the other perspective, is it a good idea to give Mary a loan, when she makes only two dollars more than John? However the bank partitions their customers in a crisp fashion, paradoxical cases like this will always be possible.

### 3.2.3 Membership

The problem with all the examples in Section 3.2.2 is that the real world is not black and white, the concepts are not crisp and everything is all in all a matter of degrees.

The translation of this fact of life into mathematical terms is the essence of fuzzy set theory. Once we recognize that shades of gray matter just as much as black and white, there is a very natural and intuitive way of generalizing the notion of a set. The key is to consider membership of an element in a set not as an all-or-nothing concept, but as a gradual attribute of elements which can be anything from "definitely not belonging" to "definitely belonging" to a set.

In formal terms, this is achieved by replacing the characteristic function with a *membership function.*

#### Definition of Fuzzy Set

Let $U$ be a universe of objects and denote by $x$ a generic element of $U$. A *fuzzy set* $A$ in $U$ is defined by a membership function $\mu_A(x)$, associating to every object in $U$ a real number in interval $[0, 1]$. The value $\mu_A(x)$ is to be understood as the degree to which $x$ belongs to $A$. If $A$ is a crisp set, $\mu_A$ reduces to the usual characteristic function of $A$.

A notational convention for fuzzy sets when the universe of discourse $U$ is discrete is, for fuzzy set $A$,

$$A = \left\{ \frac{\mu_A(x_1)}{x_1} + \frac{\mu_A(x_2)}{x_2} + \dots \right\} = \left\{ \sum_i \frac{\mu_A(x_i)}{x_i} \right\}. \tag{3.3}$$

When $U$ is continuous, the fuzzy set $A$ is denoted by

$$A = \left\{ \int_{x \in U} \frac{\mu_A(x)}{x} \right\}. \tag{3.4}$$

Both notations are nothing more than a formal device and the fractions do not have to be interpreted as divisions but just as ordered pairs, while the + does not stand for algebraic sum but rather for a function-theoretic union.

#### Membership Functions

A fuzzy set is completely defined by its membership function. Therefore, it is useful to define a few terms describing various features of this function.

Given a fuzzy set $A$, the *core* of its membership function is the (conventional) set of all elements $x \in U$ such that $\mu_A(x) = 1$. The *support* is the set of all $x \in U$ such that $\mu_A(x) > 0$.

A fuzzy set $A$ is *normal* if there is at least an $x \in U$ such that $\mu_A(x) = 1$, i.e. if its core is nonempty; it is *convex* [257] if, for any elements $x, y, z \in U$,

$$x < y < z \quad \Rightarrow \quad \mu_A(y) \geq \min\{\mu_A(x), \mu_A(z)\}, \qquad (3.5)$$

in other words, if its membership function is unimodal.

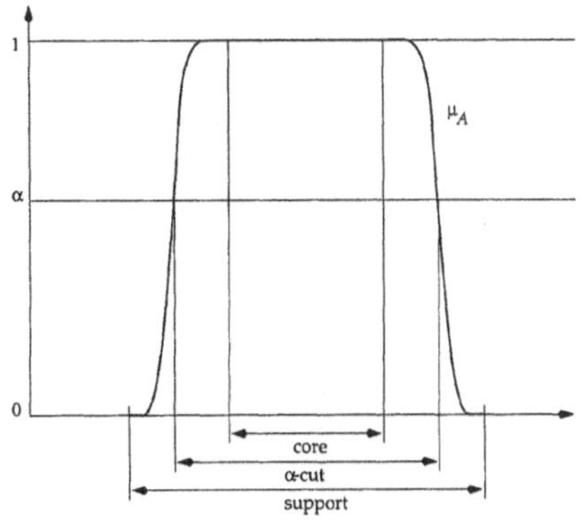


Figure 3.2 Core, support, and $\alpha$-cuts of a set $A$ of the real line, having membership function $\mu_A$.

**The Representation Theorem**

There is a nice way of deriving fuzzy set operations and properties from conventional sets. It consists in considering a fuzzy set $A$ as an infinite collection of conventional sets $\{A_\alpha\}_{0 \leq \alpha < 1}$, called $\alpha$-cuts, such that $A_\alpha = \{x : \mu_A(x) \geq \alpha\}$. Sometimes it is convenient to distinguish a *strong* $\alpha$-cut, that is a set of elements $\{x : \mu_A(x) > \alpha\}$. These sets have the property that if $\alpha \leq \beta$, $A_\beta \subseteq A_\alpha$.

The *Representation Theorem* [158] states that any fuzzy set $A$ can be represented by a union of its $\alpha$-cuts:

$$A = \bigcup_{\alpha \in [0,1]} \alpha A_\alpha. \qquad (3.6)$$

The membership value for an element $x$ will be given by the smallest $\alpha$ such that $x \in A_\alpha$.

The $\alpha$-cuts of fuzzy sets can be used to extend operations and properties of conventional sets to the new realm of fuzzy sets. For instance, given two fuzzy sets $A$ and $B$, we define their intersection $A \cap B$ as the set of intersections of all their $\alpha$-cuts: for all $0 \leq \alpha < 1$,

$$(A \cap B)_\alpha = A_\alpha \cap B_\alpha. \qquad (3.7)$$

Therefore, for all $x \in U$,

$$\begin{aligned}\mu_{A\cap B}(x) &= \min\{\alpha : x \in (A\cap B)_\alpha\} = \min\{\alpha : x \in A_\alpha \cap B_\alpha\}\\ &= \min\{\mu_A(x), \mu_B(x)\}.\end{aligned} \tag{3.8}$$

In the same way, one can derive the formula for the membership function of the union of two sets $A$ and $B$,

$$\mu_{A\cup B}(x) = \max\{\mu_A(x), \mu_B(x)\}. \tag{3.9}$$

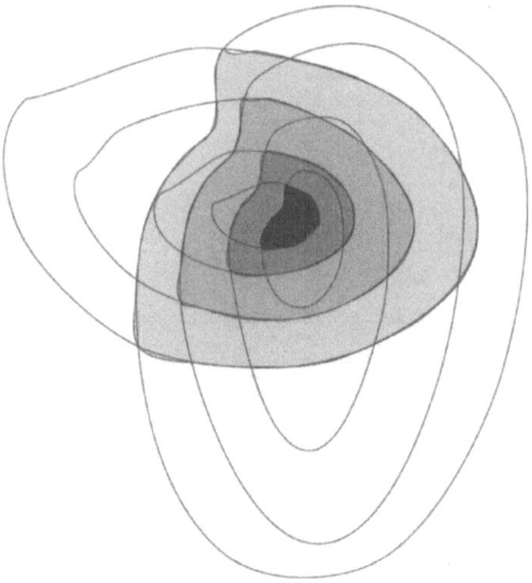

Figure 3.3 Derivation of the intersection between two fuzzy sets from the intersection between classical sets using $\alpha$-cuts.

### How Are Membership Functions Determined?

All of the above introduces fuzzy sets and shows the intimate relationship between a fuzzy set and its membership function. One question is left unanswered: who or what gives us the membership function of a fuzzy set?

From an engineering point of view, the definition of the ranges, quantities, and entities relevant to a system is a crucial design step. In fuzzy systems all entities that come into play are defined in terms of fuzzy sets, that is, of their membership functions. The determination of membership functions is then to be correctly viewed as a problem of design. As such, it can be left to the sensibility of a human expert or more objective techniques can be employed. In the former case, membership functions might be dictated by intuition, by inference from a body of facts or knowledge or by opinion methods such as a poll or an expert survey.

Alternatively, optimal membership function assignment, of course relative to a number of design goals that have to be clearly stated, such as robustness, system performance, etc., can be estimated by means of a machine learning or optimization method. In particular, neural networks and evolutionary algorithms have been employed with success to this aim. Their use in this context is of primary interest for soft computing and is described in Chapters 5 and 6 respectively.

### 3.2.4 Operations on Fuzzy Sets

The definition of operations on fuzzy sets is not unique. At an early stage of the development of fuzzy sets, two alternative expressions for union and intersection were proposed, namely, a product for intersection

$$\mu_{A \cap B}(x) = \mu_A(x)\mu_B(x), \tag{3.10}$$

and a so-called probabilistic sum for union

$$\mu_{A \cup B}(x) = \mu_A(x) + \mu_B(x) - \mu_A(x)\mu_B(x). \tag{3.11}$$

In comparison with the lattice operations defined in Equations 3.8 and 3.9, the degree of membership depends on both the values of membership function $\mu_A(x)$ and $\mu_B(x)$.

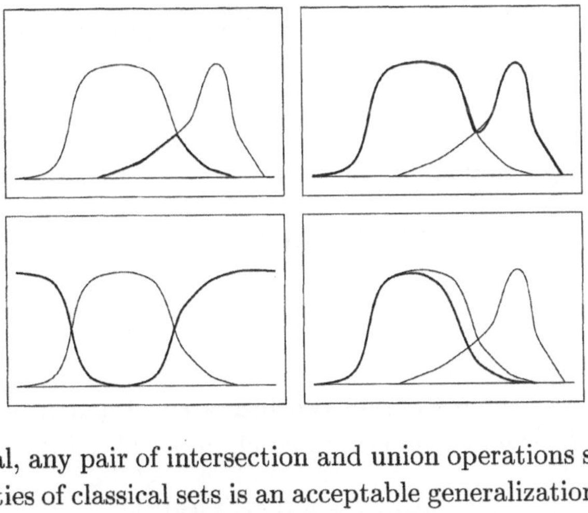

Figure 3.4 Operations on fuzzy sets. From left to right and from top to bottom: intersection, union, complement, and difference.

In general, any pair of intersection and union operations satisfying the properties of classical sets is an acceptable generalization of those operations. A broad class of models for logical connectives is formed by triangular norms and co-norms [141] (or *t-norms* and *s-norms*).

### Triangular Norms and Co-norms

A *t-norm* is a function of two arguments

$$\triangle: [0,1] \times [0,1] \to [0,1], \tag{3.12}$$

such that it is, for all $x, y, w, z \in [0,1]$,

- nondecreasing in each argument: for $x \leq y$, $w \leq z$, $x \triangle y \leq y \triangle z$;
- commutative: $x \triangle y = y \triangle x$;
- associative: $(x \triangle y) \triangle z = x \triangle (y \triangle z)$;
- satisfies the identities: $x \triangle 0 = 0$ and $x \triangle 1 = x$.

A *t-conorm* or *s-norm* is a function of two arguments

$$\triangledown: [0,1] \times [0,1] \to [0,1], \tag{3.13}$$

which is nondecreasing in each argument, commutative, associative and satisfies the identities, for $x \in [0,1]$, $x \triangledown 0 = x$, and $x \triangledown 1 = 1$.

Looking at the definitions, it is easy to verify that the properties of t-norms are the same as those of intersection and that the properties of s-norms correspond to the properties of union. An interesting fact is that the s-norm can also be derived from the relationship

$$x \triangledown y = 1 - (1-x) \triangle (1-y), \tag{3.14}$$

which happens to be another way of writing De Morgan's law.

### Union

The *union* of two fuzzy sets $A$ and $B$ is a fuzzy set $C = A \cup B$ having membership function, for all $x \in U$,

$$\mu_C(x) = \mu_A(x) \triangledown \mu_B(x), \tag{3.15}$$

where $\triangledown$ is an s-norm.

### Intersection

Similarly, the *intersection* of two fuzzy sets $A$ and $B$ is a fuzzy set $D = A \cap B$ having membership function, for all $x \in U$,

$$\mu_D(x) = \mu_A(x) \triangle \mu_B(x), \tag{3.16}$$

where $\triangle$ is a t-norm.

### Complement

The *complement* of fuzzy set $A$ is fuzzy set $\bar{A}$ having membership function, for all $x \in U$, $\mu_{\bar{A}}(x) = 1 - \mu_A(x)$.

### Arbitrariness of Operators

Having an infinite family of triangular norms gives a broad repertoire of formal models of set-theoretic connectives. The choice of a certain intersection or union operator depends on the problem to be solved. Few experiments have so far been performed as to which triangular norm best suits which type of problem; therefore, the choice is up to the sense of the "user" and subject to some trial and error.

### 3.2.5 Properties of Fuzzy Sets

It is easy to prove that fuzzy sets on universe $U$ with the union, intersection, and complement operations constitute a complete distributive lattice. Therefore, extensions of the fundamental properties of conventional sets, like De Morgan's laws, distributive laws, etc., hold.

A remarkable exception is the law of the excluded middle, which does not hold in general for fuzzy sets: in fact, for all $x \in U$, $\mu_{A \cap \bar{A}}(x) \geq 0$. In other words, the intersection of a fuzzy set with its complement can be non-empty; it is empty only in the particular case where $A$ is a conventional set.

Two fuzzy sets $A$ and $B$ are equal, $A = B$, if and only if, for all $x \in U$, $\mu_A(x) = \mu_B(x)$.

$A$ is a *fuzzy subset* of $B$ (is contained in $B$, or is less than or equal to $B$), $A \subseteq B$, if and only if, for all $x \in U$, $\mu_A(x) \leq \mu_B(x)$.

As is the case with conventional "crisp" sets, two fuzzy sets $A$ and $B$ are said to be *disjoint* if their intersection is empty, that is if, for all $x \in U$, $\mu_{A \cap B}(x) = 0$.

## 3.3 Fuzzy Relations

FUZZY SETS constitute a conceptual extension of set theory and so do fuzzy relations.

### 3.3.1 Crisp Relations

We recall that relations in classical set theory are treated as subsets of the Cartesian product of some spaces or universes.

Given $n$ spaces $U_1, U_2, \ldots U_n$, their Cartesian product (or product space) $U_1 \times U_2 \times \ldots \times U_n$ is the set of all ordered $n$-tuples $(u_1, u_2, \ldots, u_n)$, where $u_i \in U_i$ for $i = 1, 2, \ldots, n$.

A relation $R$ between spaces $U_1, U_2, \ldots U_n$ is a subset of their Cartesian product, containing all the $n$-tuples satisfying it. In symbols,

$$R \subseteq U_1 \times U_2 \times \ldots \times U_n,$$

and elements $(u_1, u_2, \ldots, u_n)$ are in relation $R$ among them if and only if $(u_1, u_2, \ldots, u_n) \in R$.

#### Properties

The properties of commutativity, associativity, distributivity, involution, and idempotency all hold for crisp relations just as they do for classical set operations, as well as De Morgan laws and the excluded middle laws. Moreover, a null relation **O** and a *complete* relation **E** can be defined acting respectively as the empty set and the universe.

#### Composition

Let $R \subseteq U \times V$ be a relation between elements from universe $U$ and elements from universe $V$, and let $S \subseteq V \times W$ be another relation relating elements from universe $V$ with elements from universe $W$. Relations $R$ and $S$ can be chained using the *composition* operations in such a way as to obtain a new relation $R \circ S \subseteq U \times W$ with $(u, w) \in R \circ S$ if and only if there exists $v \in V$ such that $(u, v) \in R$ and $(v, w) \in S$.

### 3.3.2 Fuzzy Relations

Fuzzy relations also relate elements of a number of universes to one another through the Cartesian product of the universes. However, the "strength" of the relation among the elements of an ordered $n$-tuple is a matter of degree and can vary with continuity between 0 and 1.

An $n$-ary *fuzzy relation* $R$ over universes $U_1, U_2, \ldots U_n$ is a fuzzy set over their product space, $R \subseteq U_1 \times U_2 \times \ldots \times U_n$. The membership function of $R$ is of the form $\mu_R(u_1, u_2, \ldots, u_n)$, with $u_i \in U_i$ for $i = 1, 2, \ldots, n$.

### Fuzzy Cartesian Product

Because fuzzy relations in general are fuzzy sets, we can define the fuzzy Cartesian product to be a product between two or more fuzzy sets.

Given a fuzzy set $A$ on universe $U$ and a fuzzy set $B$ on universe $V$, their Cartesian product $A \times B$ is a fuzzy relation which is contained within the full Cartesian product space $U \times V$, with membership function, for all $x \in U$ and $y \in V$,

$$\mu_{A \times B}(x, y) = \min\{\mu_A(x), \mu_B(y)\}. \tag{3.17}$$

### Composition

Let $R_1 \subseteq X \times Y$ and $R_2 \subseteq Y \times Z$ be two fuzzy relations. Their composition, $R_1 \circ R_2$, has membership function, for all $x \in X$, $z \in Z$,

$$\mu_{R_1 \circ R_2}(x, z) = \sup_{y \in Y} \min\{\mu_{R_1}(x, y), \mu_{R_2}(y, z)\}. \tag{3.18}$$

### Orthogonal Fuzzy Sets

Orthogonal fuzzy sets, also known as noninteractive fuzzy sets, can be thought of as being analogous to independent events in probability theory. They always arise in the context of relations or in $n$-dimensional mappings.

Let $A$ be a fuzzy set defined on product space $U = X \times Y$. The set $A$ is separable into two *noninteractive* fuzzy sets, called its orthogonal projections, if and only if

$$A = \mathrm{proj}_X(A) \times \mathrm{proj}_Y(A), \tag{3.19}$$

where $\mathrm{proj}_X(A)$ is such as, for all $x \in X$,

$$\mu_{\mathrm{proj}_X(A)}(x) = \sup_{y \in Y} \mu_A(x, y), \tag{3.20}$$

and $\mathrm{proj}_Y(A)$ is such as, for all $y \in Y$,

$$\mu_{\mathrm{proj}_Y(A)}(x) = \sup_{x \in X} \mu_A(x, y). \tag{3.21}$$

Orthogonality means that $A$ can be uniquely reconstructed by its projections.

### 3.3.3 Fuzzy Equivalence Relations

An equivalence relation $R$ is a binary relation defined between elements of the same set, say $X$, having the following three properties:

1. $x\,R\,x$ for all $x \in X$ (reflexive);
2. $x\,R\,y$ implies $y\,R\,x$ for all $x, y \in X$ (symmetric);
3. $x\,R\,y$ and $y\,R\,z$ imply $x\,R\,z$ for all $x, y, z \in X$ (transitive).

The fuzzy counterparts of equivalence relations are also known as *similarity measures*, in that they relate elements that may be just similar, not only perfectly equivalent. The membership function of a fuzzy equivalence relation $R \subseteq X \times X$ must satisfy the following three properties:

1. $\mu_R(x, x) = 1$ for all $x \in X$ (reflexive);
2. $\mu_R(x, y) = \mu_R(y, x)$ for all $x, y \in X$ (symmetric);
3. $\mu_R(x, z) = \min\{\mu_R(x, y), \mu_R(y, z)\}$, for all $x, y, z \in X$ (transitive).

The relation "$x$ is similar to $y$", where $x$ and $y$ are two real numbers is an example of a fuzzy equivalence relation. Its membership function might be of the form

$$\mu_{\text{similar}}(x, y) = \frac{1}{1 + (x - y)^2}.$$

Of course, fuzzy equivalence relations can be defined between objects of any kind, like pictures, texts, personal records, sounds, and so on.

## 3.4 The Extension Principle

The *Extension Principle* is the main formal tool for making any mathematical theory fuzzy in a consistent and well-founded way. As a matter of fact, all mathematical sciences might be reformulated using the new mathematical language stemming from fuzzy set theory. The way this can be done is explained in this section, while Section 3.5 provides one example of the application of the method, namely to arithmetic.

Let $U$ be the Cartesian product of $n$ universes $U_1, \ldots, U_n$ and let $A_1, \ldots, A_n$ be an equal number of fuzzy sets defined in $U_1, \ldots, U_n$ respectively.

Suppose $t: U \rightarrow V$ is a morphism from $U$ into a new universe $V$. The question we ask is what the image of a fuzzy subset of $U$ in this new universe $V$ would be under the morphism $t$. This image would also be a fuzzy set, and its membership function would be calculated from the memberhip function of the original set and the morphism $t$.

Let $B$ represent the fuzzy set induced in $V$ by morphism $t$ from the fuzzy sets $A_1, \ldots, A_n$ defined in $U$. The Extension Principle states that $B$ has membership function, for all $y \in V$,

$$\mu_B(y) = \sup_{(x_1,\ldots,x_n)\in t^{-1}(y)} \min\{\mu_{A_1}(x_1), \ldots, \mu_{A_n}(x_n)\}. \tag{3.22}$$

$B$ is said to extend fuzzy sets $A_1, \ldots, A_n$ in $V$.

Equation 3.22 is expressed for morphisms $t$ of general form. If $t$ is a discrete-valued function, the sup operator can be replaced by the max operator.

## 3.5 Fuzzy Arithmetic

THIS SECTION introduces the concept of *fuzzy number* and develops the fuzzification of arithmetic, using the Extension Principle illustrated in Section 3.4.

### 3.5.1 Fuzzy Numbers

Fuzzy numbers are just a special type of fuzzy sets, whose universe of discourse is the set of real numbers. Fuzzy numbers can be conveniently defined as a generalization of an interval of confidence [111], taking into account different *levels of presumption*. The maximum of presumption is considered to be at level 1 and the minimum of presumption to be at level 0. The level of presumption $\alpha \in [0,1]$ gives an interval of confidence $A_\alpha = [a_1^{(\alpha)}, a_2^{(\alpha)}]$, which is a monotonic decreasing function of $\alpha$, that is,

$$\alpha' > \alpha \Rightarrow A_{\alpha'} \subset A_\alpha. \tag{3.23}$$

Now, these intervals of confidence are nothing but alpha cuts of a fuzzy set $A$. Accordingly, we can define a fuzzy number $A$ as a fuzzy set defined on $\mathbb{R}$ such that

1. $A$ is normal, i.e. there exists at least one $x \in \mathbb{R}$ for which $\mu_A(x) = 1$;
2. $A$ is convex, i.e. $\mu_A$ is unimodal;
3. $A$ is upper semicontinuous;
4. $A$ has a bounded support.

By recalling the representation theorem, 2 and 3 can be replaced by the two equivalent conditions

- all $\alpha$-cuts of $A$ are convex;
- all $\alpha$-cuts of $A$ are closed intervals of $\mathbb{R}$.

Fuzzy numbers model approximate numerical notions such as, for example, *near* zero, *about* five, and so on. Unimodality of $\mu_A$ ensures that there exist only one interval in which the relevant elements have the highest degree of membership.

Fuzzy arithmetic studies the properties and operations of fuzzy numbers, much like conventional arithmetic studies the properties and operations of crisp numbers.

### 3.5.2 Operations on Fuzzy Numbers

Operations on fuzzy numbers are defined by applying the extension principle on the basis of operations on crisp numbers. They are, therefore, generalizations of the conventional operations and entail them as special cases when the operands are crisp.

In general, the fuzzy counterpart $F$ of a conventional function $f: \mathbb{R}^n \to \mathbb{R}$ is defined as follows, for all $A_1, \ldots, A_n$ fuzzy numbers and $z \in \mathbb{R}$:

$$\mu_{F(A_1,\ldots,A_n)}(z) = \sup_{\substack{x_1,\ldots,x_n \in \mathbb{R} \\ z = f(x_1,\ldots,x_n)}} \inf\{\mu_{A_1}(x_1), \ldots, \mu_{A_n}(x_n)\}. \qquad (3.24)$$

This formula assumes using min as the t-norm and max as the s-norm, but it can be easily generalized to any other pair of triangular norm and co-norm by replacing the sup and inf operators with the appropriate fuzzy union and intersection operators. A sufficient condition that guarantees $F(A_1, \dots, A_n)$ to be in turn a fuzzy number is given in [55].

The four elementary operations on fuzzy numbers can now be defined with very little effort, for all $x, y, z \in \mathbb{R}$:

$$\mu_{A+B}(z) = \sup_{x+y=z} \min\{\mu_A(x), \mu_B(y)\}, \tag{3.25}$$

$$\mu_{A-B}(z) = \sup_{x-y=z} \min\{\mu_A(x), \mu_B(y)\}, \tag{3.26}$$

$$\mu_{AB}(z) = \sup_{xy=z} \min\{\mu_A(x), \mu_B(y)\}, \tag{3.27}$$

$$\mu_{A/B}(z) = \sup_{x/y=z} \min\{\mu_A(x), \mu_B(y)\}. \tag{3.28}$$

Figure 3.5 shows examples of operations on fuzzy numbers. Note that all four operations require the solution of a mathematical optimization problem with constraints resulting from the corresponding operator.

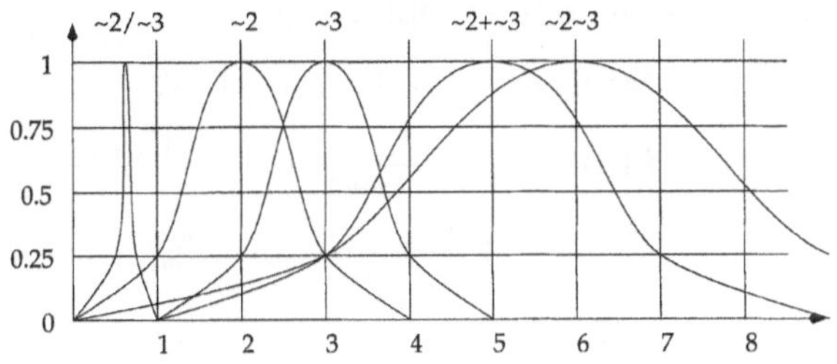


Figure 3.5 Sum, product and division of two fuzzy numbers, $\sim 2$ and $\sim 3$.

### 3.5.3 Families of Fuzzy Numbers

Calculating the result of an operation on fuzzy numbers amounts to solving an optimization problem. Therefore, if no simplifying assumption is made, the computational cost of fuzzy arithmetic can be overwhelming, to the extent of making even simple arithmetical problems intractable.

This is the cost of adopting a more realistic notion of number. Are we doomed to frustration or is there a way out? Fortunately, the answer is yes, there is a way out, provided we restrict ourselves to fuzzy numbers of specific forms.

In order to be a family, a specific form of fuzzy numbers must be closed with respect to the arithmetic operations.

If we are able to define a family of fuzzy numbers having membership functions completely determined by a few parameters, we can redefine arithmetic operations on fuzzy numbers as operations on those parameters. The resulting parameters, in turn, will define a fuzzy number—the result of an operation. By doing this, we would have a simple and effective way of making calculations.

One such family of fuzzy numbers are the so-called L-R fuzzy numbers (L-R stands for left- and right-side). An L-R fuzzy number is defined by membership function

$$\mu(x) = \begin{cases} L\left(\frac{m-x}{\lambda}\right), & \text{if } x \leq m, \\ R\left(\frac{m-x}{\rho}\right), & \text{if } x \geq m, \end{cases} \tag{3.29}$$

where $L$ and $R$ are two functions enjoying the following properties:

1. they are symmetric around zero: $L(-x) = L(x)$;
2. they have their mode in zero: $L(0) = 1$;
3. they are decreasing in $[0, +\infty]$.

Examples of such functions are, for $p > 0$:

$$\begin{aligned} L(x) &= \max\{0, (1 - |x|)^p\}, \\ L(x) &= \frac{1}{1 + |x|^p}, \\ L(x) &= e^{-\frac{x^2}{p}}, \\ &\dots \end{aligned}$$

In Equation 3.29, $\lambda$ and $\rho$ are parameters controlling the degree of fuzziness of the fuzzy number. By using membership functions of the form of Equation 3.29, a fuzzy number will be denoted by $(m, \lambda, \rho)$ and all basic arithmetic operations will be defined in terms of this parametric representation, for example

$$\begin{aligned} (m_1, \lambda_1, \rho_1) + (m_2, \lambda_2, \rho_2) &= (m_1 + m_2, \lambda_1 + \lambda_2, \rho_1 + \rho_2), \\ (m_1, \lambda_1, \rho_1) - (m_2, \lambda_2, \rho_2) &= (m_1 - m_2, \lambda_1 + \lambda_2, \rho_1 + \rho_2). \end{aligned}$$

The relevant formulas for multiplication and division are more complicated and in some cases they only give approximate results. In fact, it is easy to show that the multiplication of L-R fuzzy numbers does not give an L-R fuzzy number except for particular cases.

A particular case of L-R fuzzy numbers are so-called *triangular* fuzzy numbers.

### 3.5.4 Imprecision Propagation

The imprecision inherent in a fuzzy number tends to increase if we obtain it as a result of operations on other fuzzy numbers. Since all operations, without exception, amplify imprecision, the result of any sufficiently long chain of operations will be a maximally imprecise number. This effect should not come as a surprise to the reader acquainted with numerical analysis, where it is studied by the theory of error propagation.

This should tell us something about the presumptions that are implicitly made when "pretending" that the data of a problem are known with precision and about the validity of the solutions thus obtained.

## 3.6 Fuzzy Logic

THE PHRASE *fuzzy logic* has two readings. In the broad sense it is used to dub the discipline that deals with anything fuzzy, like sets, predicates, values, etc. In the narrow sense, it is the name of a particular type of multivalued logic, dealing with uncertainty and partial truth, having fuzzy set theory as its foundation. This chapter is intended to provide an introduction to fuzzy logic in this narrow sense.

One of the most natural and common definitions of logic presents this discipline as the analysis of reasoning methods. When studying these methods, logic focuses on the form rather than the contents of an argument: the truth or falsity of the particular premises or conclusions do not matter, but whether the truth of premises implies the truth of conclusions matters.

The basic objects of logic are *propositions* which have a truth value, like, for instance, "today it is raining". When propositions express the property of objects that can be replaced by variables, they are referred to as *predicates*, like "... is a nice person", where the dots can stand for the name of any person.

### 3.6.1 Sets, Propositions, and Predicates

A one-to-one mapping can be established between the realm of (fuzzy) sets and the realm of (fuzzy) logic propositions. In classical logic, a given proposition can fall in either of two sets: the set of all true propositions and the set of all false propositions, which is the complement of the former. In fuzzy logic, the set of true proposition and its complement, the set of false propositions, are fuzzy. The degree to which a given proposition belongs to the set of true propositions is its degree of truth.

Much in the same way, a one-to-one mapping can be established as well between (fuzzy) sets and (fuzzy) predicates. In classical logic, a predicate of an element of the universe of discourse defines the set of elements for which that predicate is true and its complement, the set of elements for which that predicate is not true. Once again, in fuzzy logic, these sets are fuzzy and the degree of truth of a predicate of an element is given by the degree to which that element is in the set associated with that predicate.

Given a fuzzy set $A$ in a universe of discourse $X$, the *fuzzy predicate* $P_A$ associated with it can be defined, whose truth value on an element $x \in X$ is given by $\mu_A(x)$. Predicate $P_A(x)$ may also be written in the form $x \in A$ or $x$ is $A$.

A logical system is obtained in this way, known as *Fuzzy Logic*. Its goal is to formalize the mechanisms of approximate reasoning. As such, it has its own place in the domain of multivalued logics, although its motivations significantly differ from other systems of the same kind, as, for instance, Łukasiewicz's logics.

It is worth mentioning that many of the concepts which contribute to the success of fuzzy logic as logics of approximate reasoning are not part of more traditional many-valued logic systems. These include the concepts of *linguistic variable, canonical form, fuzzy rule* and *fuzzy quantifier*, as well as inference methods such as *interpolative, syllogistic,* and *dispositional* reasoning.

### 3.6.2 Logical Connectives

A fuzzy logic proposition $P$ is a statement involving some concept without clearly defined boundaries. Linguistic statements that tend to express subjective ideas and that can be interpreted slightly differently by various individuals typically involve fuzzy propositions. Natural language is mostly fuzzy, in that it involves vague and imprecise terms. The truth value assigned to $P$ can be any value in the interval $[0, 1]$, where zero represents complete falsity and one complete truth. The assignment of a truth value to a proposition is a mapping from the set of all propositions $\mathcal{P}$ to the interval $[0, 1]$:

$$T: \mathcal{P} \rightarrow [0, 1]. \tag{3.30}$$

Function $T$ is called the *truth* function.

The logical connectives of negation, disjunction, and conjunction can be defined for fuzzy logic based on its set-theoretic foundation, as follows:

$$\text{Negation} \quad T(\neg P) = 1 - T(P); \tag{3.31}$$

$$\text{Disjunction} \quad T(P \vee Q) = \max\{T(P), T(Q)\}; \tag{3.32}$$

$$\text{Conjunction} \quad T(P \wedge Q) = \min\{T(P), T(Q)\}. \tag{3.33}$$

In the above definitions, of course, the min and max operators can be replaced by any suitable pair of triangular norm and co-norm.

### 3.6.3 Fuzzy Implication and Inference

When it comes to implication, unfortunately, things are not as linear as for the other logical connectives, as many definitions are possible depending on the semantic interpretation or on the context in which implication is used.

In classical logic, implication is defined in terms of the other connectives as

$$P \supset Q \quad \equiv \quad \neg P \vee Q. \tag{3.34}$$

If we transpose this formula in fuzzy logic, we get one definition of implication [258] whereby

$$T(P \supset Q) = \max\{1 - T(P), T(Q)\}. \tag{3.35}$$

The implication connective can also be modeled in rule-based form, that is

$$P_A \supset P_B \quad \equiv \quad \text{if } x \in A \text{ then } y \in B, \tag{3.36}$$

which is equivalent to a fuzzy relation $R = (A \times B) \cup (\bar{A} \times U)$, where $U$ is the universe of discourse in which $B$ is defined. The membership function of relation $R$ is expressed by the formula

$$\mu_R(x, y) = \max\{\min[\mu_A(x), \mu_B(y)], 1 - \mu_A(x)\}, \quad (3.37)$$

whence one can obtain an alternative definition for the truth of implication,

$$T(P \supset Q) = \max\{\min[T(P), T(Q)], 1 - T(P)\}. \quad (3.38)$$

There are other techniques for obtaining the fuzzy relation $R$ based on the rule "if $x \in A$ then $y \in B$", from which the truth of the compound proposition $P_A \supset P_B$ can be derived.

A very popular definition of fuzzy implication, which might be referred to as Mamdani's implication, after Prof. Mamdani's work on fuzzy control [132], is

$$\mu_R(x, y) = \min\{\mu_A(x), \mu_B(y)\}, \quad (3.39)$$

which is equivalent to the fuzzy Cartesian product of fuzzy sets $A$ and $B$, i.e. $R = A \times B$.

Another formulation for the implication, known as Łukasiewicz' implication after the Polish logician Jan Łukasiewicz [187], is the following:

$$\mu_R(x, y) = \min\{1, 1 - \mu_A(x) + \mu_B(y)\}. \quad (3.40)$$

A slight variation of this formulation,

$$\mu_R(x, y) = \min\{1, \mu_A(x) + \mu_B(y)\}, \quad (3.41)$$

is known as the *bounded sum* implication.

A definition of fuzzy implication due to Goguen [75] is, for all $x$ such that $\mu_A(x) > 0$,

$$\mu_R(x, y) = \min\{1, \frac{\mu_B(y)}{\mu_A(x)}\}. \quad (3.42)$$

Two forms of *correlation-product* implication are based on the notions of conditioning [227],

$$\mu_R(x, y) = \max\{\mu_A(x)\mu_B(y), 1 - \mu_A(x)\}, \quad (3.43)$$

and reinforcement

$$\mu_R(x, y) = \mu_A(x)\mu_B(y). \quad (3.44)$$

Both of these product forms tend to dilute the influence of joint membership values that are small and, as such, are related to Hebbian-type learning algorithms (cf. Chapter 2, Section 2.6.1).

So-called Brouwerian implication, discussed in [198] is defined as

$$\mu_R(x,y) = \begin{cases} 1, & \text{if } \mu_A(x) \leq \mu_B(y), \\ \mu_B(y), & \text{otherwise.} \end{cases} \tag{3.45}$$

A simplified form of the above, giving only bipartite truth values,

$$\mu_R(x,y) = \begin{cases} 1, & \text{if } \mu_A(x) \leq \mu_B(y), \\ 0, & \text{otherwise.} \end{cases} \tag{3.46}$$

has been termed in the literature as *R-SEQ (standard sequence logic* implication [137].

Some of the implications illustrated in this section are more broadly used than others, while some are useful in particular domains or situations. The bottom line is that the appropriate choice of an implication operator depends on the context of an application and is a matter left to the analyst.

### 3.6.4 Approximate Reasoning

Much of human reasoning is approximate rather than precise in nature; indeed one could argue that only a small fraction of our thinking can be classified as precise in logical or quantitative terms.

The ultimate goal of fuzzy logic is to provide the theoretical foundation for reasoning about imprecise propositions. Such a task has been referred to as *approximate* reasoning [256]. Approximate reasoning studies chains of inferences in fuzzy logic, thus being analogous to predicate logic for reasoning with precise propositions, and hence can be regarded as an extension of classical propositional calculus that deals with partial truths.

An elementary example of approximate reasoning is the following variation of the famous Aristotelian syllogism,

$$\begin{array}{rcl} P_1 & : & \textit{Most} \text{ men are } \textit{vain} \\ P_2 & : & \text{Socrates is a man} \\ \hline P_3 & : & \text{It is } \textit{likely} \text{ that Socrates is } \textit{vain}, \end{array} \tag{3.47}$$

which can be defined in terms of fuzzy composition (see Section 3.3).

It is useful to think of reasoning as the process of solving a system of relational assignment equations.

Consider proposition $P_A \equiv x \in A$. We can interpret $x \in A$ as the assignment of a unary fuzzy relation $A$ to a variable corresponding to an implied attribute $R$ of $x$:

$$R(x) = A. \tag{3.48}$$

Similarly, the compound proposition $Q \equiv (x \in A) \wedge (y \in B)$ is equivalent to two assignment equations

$$R_1(x) = A, \tag{3.49}$$

$$R_2(y) = B. \tag{3.50}$$

A system of relational assignment equations like these is solved by calculating the composition of the relevant relations, in this case $R = R_1 \circ R_2 = R_1 \times R_2$, because both relations are unary.

Imagine instead having an implication $P_A \supset P_B$, which can be modeled with the *binary* fuzzy relation $R$ based on the rule "if $x \in A$ then $y \in B$". Furthermore, suppose that a new antecedent proposition, say $P_{A'}$, is given, equivalent to $x \in A'$. Composition allows us to derive a consequent proposition $P_{B'} \equiv y \in B'$, much in the same way as in classical propositional calculus one can apply *modus ponens* to the two propositions $P \supset Q$ and $P$ to derive $Q$. The fuzzy counterpart of *modus ponens* works by calculating

$$B' = A' \circ R.. \tag{3.51}$$

One might wonder whether composing $R$ with the original antecedent $A$ would yield the original consequent $B$. In fact, this is not the case: in general $A \circ R \neq B$.

### 3.6.5 Fuzzy Rule-Based Systems

#### Linguistic Variables

A linguistic variable [259] is defined on a numerical interval and has linguistic values, whose semantics is defined by their membership function. For example, a linguistic variable *temperature* might be defined over the interval $[-20°C, 50°C]$; it could have linguistic values like *cold, warm,* and *hot,* whose meanings would be defined by appropriate membership functions.

## Fuzzy Rules

The concept of *fuzzy rule* is strictly related to the concept of fuzzy relation (see Section 3.3). A fuzzy rule is a syntactic structure of the form

$$\text{IF } \textit{antecedent} \text{ THEN } \textit{consequent}, \tag{3.52}$$

where each *antecedent* and *consequent* are well-formed formulas in fuzzy logic.

Fuzzy rules provide an alternative, compact, and powerful way of expressing functional dependencies between various elements of a system in a modular and, most importantly, intuitive fashion. As such, they have found broad application in practice, for example in the field of control and diagnostic systems.

## Inference in Fuzzy Rule-Based Systems

The semantics of a fuzzy rule-based system is governed by the laws of the calculus of fuzzy rules [261]. In summary, all rules in a fuzzy rule base take part simultaneously in the inference process, each to an extent proportionate to the truth value associated with its antecedent. The result of an inference is represented by a fuzzy set for each of the dependent variables. The degree of membership for a value of a dependent variable in the associated fuzzy set gives a measure of its compatibility with the observed values of the independent variables.

Given a system with $n$ independent variables $x_1, \ldots, x_n$ and $m$ dependent variables $y_1, \ldots, y_m$, let $R$ be a base of $r$ fuzzy rules

$$\begin{array}{ll} \text{IF } P_1(x_1, \ldots, x_n) & \text{THEN } Q_1(y_1, \ldots, y_m), \\ \vdots & \vdots \\ \text{IF } P_r(x_1, \ldots, x_n) & \text{THEN } Q_r(y_1, \ldots, y_m), \end{array} \tag{3.53}$$

where $P_1, \ldots, P_r$ and $Q_1, \ldots Q_r$ represent fuzzy predicates respectively on independent and dependent variables, and let $\tau_P$ denote the truth value of predicate $P$. Then the membership function describing the fuzzy set of values taken up by dependent variables $y_1, \ldots, y_m$ of system $R$ is given by

$$\begin{aligned} &\tau_R(y_1, \ldots, y_m; x_1, \ldots, x_n) \\ &\quad = \sup_{1 \leq i \leq r} \min\{\tau_{Q_i}(y_1, \ldots, y_m), \tau_{P_i}(x_1, \ldots, x_n)\}. \end{aligned} \tag{3.54}$$

## The Mamdani Model

The type of fuzzy rule-based system just described, making use of the min and max as the triangular norm and co-norm, is called the

Mamdani model. A Mamdani system has rules of the form

$$\text{IF } x_1 \text{ is } A_1 \text{ AND} \ldots \text{AND } x_n \text{ is } A_n \text{ THEN } y \text{ is } B, \qquad (3.55)$$

where the $A_i$s and $B$ are linguistic values (i.e., fuzzy sets) and each clause of the form "$x$ is $A$" has the meaning that the value of variable $x$ is in fuzzy set $A$.

#### The Sugeno Model

The Sugeno model, also known as Takagi-Sugeno-Kang (TSK) model [216], is a variant of the Mamdani model. It uses rules of the form

$$\text{IF } x_1 \text{ is } A_1 \text{ AND} \ldots \text{AND } x_n \text{ is } A_n \text{ THEN } y = f(x_1, \ldots, x_n), \qquad (3.56)$$

where $f$ can be in principle any function of the input variables taking values in the output variable range. Usually, however, $f$ is restricted to being a linear combination of the input variables,

$$f(x_1, \ldots, x_n) = w_0 + w_1 x_1 + \ldots + w_n x_n,$$

with $w_0$, $w_1$, $\ldots$, $w_n$ real constants that are part of the rule specification. Min or product may be used to perform conjunction.

The combined result of applying the rules of a Sugeno-type fuzzy system is a crisp number, computed as the average of the outputs of the single rules, weighted by the degrees of truth of their antecedents. This is just a particular case of the weighted average method of defuzzification illustrated below.

### 3.6.6 Defuzzification Methods

There may be situations in which the output of a fuzzy inference needs to be a crisp number $y^*$ instead of a fuzzy set $R$. Defuzzification is the conversion of a fuzzy quantity into a precise quantity.

At least seven methods in the literature are popular for defuzzifying fuzzy outputs [88].

#### Height Method

Also known as the *maximum membership* principle, this method consists in choosing the number with maximal degree of membership. Of course, the applicability of this method is limited to peaked membership functions.

### Centroid Method

This method, also called *Center of Area* or *Center of Gravity*, is the most prominent and physically appealing of all the defuzzification methods. It results in a crisp value

$$y^* = \frac{\int y\mu_R(y)dy}{\int \mu_R(y)dy}, \tag{3.57}$$

where the integration can be replaced by summation in discrete cases.

### Weighted Average Method

This method is only valid for an output membership function composed by a number of symmetric membership functions $\mu_i$. It weights each membership function in the output by its respective maximum membership value (the mode):

$$y^* = \frac{\sum_i \bar{y}_i \mu_i(\bar{y}_i)}{\sum_i \mu_i(\bar{y}_i)}, \tag{3.58}$$

where $\bar{y}_i$ represents the mode of membership function $\mu_i$.

### Mean-Max Method

This method, also called *middle of maxima*, is a slight generalization of the height method to the case where there is more than one value with maximum degree of membership. It takes as $y^*$ the midpoint between the smallest and the largest number having maximum degree of membership:

$$y^* = \frac{m+M}{2}, \tag{3.59}$$

where $m$ and $M$ are respectively the minimum and the maximum $y$ such that, for all $z$, $\mu_R(y) \geq \mu_R(z)$.

### Center of Sums

This method is faster than most other methods in use. It involves the algebraic sum of the integrals of the single membership functions that make up the output membership function, instead of the integral of their union. By doing this, however, the intersecting areas are added twice, introducing a bias. The defuzzified value is given by the formula

$$y^* = \frac{\int y \sum_i \mu_i(y)dy}{\int \sum_i \mu_i(y)dy}. \tag{3.60}$$

This method is similar to the weighted average method, but here the weights are the areas (i.e. integrals) of the respective membership functions, instead of the individual membership values.

#### Center of Largest Area

If the output fuzzy set has at least two convex subregions, then the center of gravity (i.e. the $y^*$ calculated with the the centroid method) of the convex subregion with the largest area is used as $y^*$.

#### First (or Last) of Maxima

This method is another generalization of the height method for the case in which the output membership function has more than one maximum. Then either the first or the last (depending on the application) of the maxima is used as the defuzzified value.

## 3.7 Possibility Theory

POSSIBILITY theory, and the related belief and evidence theory, provide a way of treating uncertain information about events that is alternative and complementary to probability theory. All these theories, including probability theory, are mathematically related and all are used to model various forms of uncertainty. As more information about a problem becomes available, the mathematical description of uncertainty can easily transform from one theory to the next in a sort of spectrum of precision.

### 3.7.1 Measures

Possibility, belief, and probability theory are all founded on the mathematical concept of *measure*.

A *measurable space* is any ordered pair $(U, \mathcal{F})$, where $U$ is a set and $\mathcal{F}$ is a $\sigma$-algebra of parts of $U$. In practice, if $U$ is finite, $\mathcal{F} = 2^U$, the entire power set of $U$.

A *measure* on a measurable space $(U, \mathcal{F})$ is a function $m: \mathcal{F} \to \mathbb{R}^+$ such that

1. $m(\emptyset) = 0$;
2. $m(F) \geq 0$, for all $F \in \mathcal{F}$;
3. $m(F) \leq m(G)$, for all $F, G \in \mathcal{F}$ such that $F \subseteq G$.

A remarkable property enjoyed by some popular types of measures (e.g., probability) is *additivity*: for all numerable sequences of mutually disjoint sets $F_1, F_2, \ldots \in \mathcal{F}$,

$$m\left(\bigcup_{i=1}^{\infty} F_i\right) = \sum_{i=1}^{\infty} m(F_i).$$

The interested reader can find a more detailed account of the mathematical theory of fuzzy measures in [113], just to cite one example.

### 3.7.2 Possibility Distributions

Fuzzy sets allow the representation of imprecise information. Information is imprecise when the value of the variable to which it refers cannot be completely determined within a given universe of discourse.

For example, the construction of a building can be decomposed in a number of tasks whose individual duration is not known. We only know of each task that it might not be carried out in less than a minimum time and that it should be completed in less than a maximum time. Given just this information, the time needed for the construction cannot be estimated *a priori*, nor can a probability be given for it.

The membership function of a fuzzy set describing imprecise information, that is, the more or less plausible and mutually exclusive values for one or more variables, such as construction time in the above example, are to be viewed as a *possibility distribution* [260].

If $A$ is the fuzzy set of values that a variable $x$ can take up, we denote by $\Pi_x = \mu_A$ the possibility distribution attached to $x$. The identity $\Pi_x(u) = \mu_A(u)$ means that the degree to which value $u$ belongs to $A$ is the same as the possibility degree of $x$ being equal to $u$ when all is known about $x$ is that it is constrained to take a value in $A$.

A possibility distribution for which there exists a completely possible value is said to be *normalized*.

### 3.7.3 Possibility and Necessity Measures

A possibility distribution $\Pi$ induces a *possibility measure* and its dual *necessity measure*, denoted by $\pi$ and $\eta$ respectively. Both measures apply to a crisp set $A$ and are defined as follows:

$$\pi(A) \equiv \sup_{s \in A} \Pi(s); \tag{3.61}$$

$$\eta(A) \equiv 1 - \pi(\bar{A}) = \inf_{s \in \bar{A}} \{1 - \Pi(s)\}. \tag{3.62}$$

In words, the possibility measure of set $A$ corresponds to the greatest of the possibilities associated to its elements; conversely, the necessity measure of $A$ is equivalent to the impossibility of its complement $\bar{A}$.

A few properties of possibility and necessity measures induced by a normalized possibility distribution on a finite universe of discourse $U$ are the following, for all subsets $A, B \subseteq U$:

- $\pi(A \cup B) = \max\{\pi(A), \pi(B)\}$;
- $\pi(\emptyset) = 0$, $\pi(U) = 1$;
- $\eta(A \cap B) = \min\{\eta(A), \eta(B)\}$;
- $\eta(\emptyset) = 0$, $\eta(U) = 1$;
- $\pi(A) = 1 - \eta(\bar{A})$ (duality);
- $\pi(A) \geq \eta(A)$;
- $\eta(A) > 0$ implies $\pi(A) = 1$;
- $\pi(A) < 1$ implies $\eta(A) = 0$.

An immediate consequence of these properties is that either $\pi(A) = 1$ or $\pi(\bar{A}) = 1$. Both a set $A$ and its complement having a possibility of 1 is the case of complete ignorance on $A$.

The above definitions of $\pi$ and $\eta$, given in terms of a crisp set $A$, can be extended to fuzzy sets as follows:

$$\pi(A) \equiv \sup_{u \in U} \min\{\mu_A(u), \Pi(u)\}; \tag{3.63}$$

$$\eta(A) \equiv 1 - \pi(A) = \inf_{u \in U} \max\{\mu_A(u), 1 - \Pi(u)\}. \tag{3.64}$$

### 3.7.4 Belief and Plausibility Measures

A *belief measure* is a mapping, denoted bel($A$), that associates the degree of support, or evidence, to an event $A$, i.e. a subset of the universe of discourse $U$. The *plausibility measure* of an event $A$ is defined as

$$\mathrm{pl}(A) \equiv 1 - \mathrm{bel}(\bar{A}). \tag{3.65}$$

In words, we might say that an event is plausible to the extent that there is no evidence supporting its contrary. This duality between belief and plausibility is formally identical to the duality between possibility and necessity.

By convention, $\mathrm{bel}(U) = 1$, i.e. there is total belief that the actual value of an uncertain variable is to be found in its universe of discourse; furthermore, since $\mathrm{bel}(\emptyset) = 0$, $\mathrm{pl}(U) = 1$ as well. However, it is entirely possible that, for a given $A \subset U$, $\mathrm{bel}(A) + \mathrm{bel}(\bar{A}) < 1$; in other words belief need not be additive. When additivity holds, the belief measure is in fact a basic probability assignment and the evidence about subsets of $U$ can be described probabilistically. The difference $1 - \mathrm{bel}(A) + \mathrm{bel}(\bar{A})$ is called the *ignorance* about $A$.

If given, basic probability assignment or probability measure $m$ can be used to determine a belief measure. For all crisp sets $A \subset U$, we define

$$\mathrm{bel}(A) = \sum_{B \subseteq A} m(B). \tag{3.66}$$

While $m(A)$ is the evidence supporting just the set $A$, according to this definition $\mathrm{bel}(A)$ is the overall evidence supporting set $A$ and all its subsets.

In a similar way, the same measure $m$ can induce the plausibility measure

$$\mathrm{pl}(A) = \sum_{B \cap A \neq \emptyset} m(B), \tag{3.67}$$

where the plausibility of $A$ amounts to the total evidence supporting all sets in the universe of discourse that have a nonempty intersection with $A$, including proper subsets of $A$.

The mathematical study of the relationship between probability, belief, and plausibility forms the object of the Dempster-Shafer theory of evidence [207], which is beyond the scope of this introduction.

## 3.8 Applications of Fuzzy Systems

APPLICATIONS of fuzzy systems are very numerous. In 1999, ERUDIT, the European network of excellence in fuzzy logic, issued its "technological roadmap", containing an interesting taxonomy of applications, classified by their industrial or business sector. The list is the following:

1. Primary and Process Industry, including measurements and signal processing, process control (embedded and supervisory), fault diagnosis, quality control, planning, and scheduling;

2. Components, Capital, and Consumer Goods;

3. Traffic, including data analysis and control;

4. Telecommunication, including call admission control, parameter estimation, forecasting, routing, traffic policing, usage parameter control, ATM traffic shaping and flow control, network traffic modeling, network management, and telecommunication services;

5. Human, Medicine, and Healthcare, divided in conservative disciplines, invasive medicine, regionally defined medical disciplines, neuromedicine, medical image and signal processing, laboratory, basic medical science, nursing, healthcare, and other application domains;

6. Finance, Trade, and Services, including risk management, forecasting, customer segmentation, fraud detection, customer-oriented marketing;

7. Cross Sectional Areas:

    (a) Nuclear Engineering, including control in and of nuclear power plants, safety management, accounting of nuclear waste, nuclear energy, and public opinion;

    (b) Ecological Engineering, including control of technology, data analysis, ecological modeling, and forecasting.

By reading this list, which leaves out at least the defense domain, one would be correctly led to think that fuzzy systems can be, and have been, applied to almost everything.

Due to obvious space constraints, only an overview of the most prominent application domains can be given here.

### 3.8.1 Fuzzy Control

The purpose of control is to influence the behavior of a system by acting on some input variables of that system according to a set of rules that model how the system operates. This is sketched in Figure 3.6.

Classic control theory uses a mathematical model to define a relationship that transforms the desired state and the observed state of the system into inputs that will alter the future state of that system.

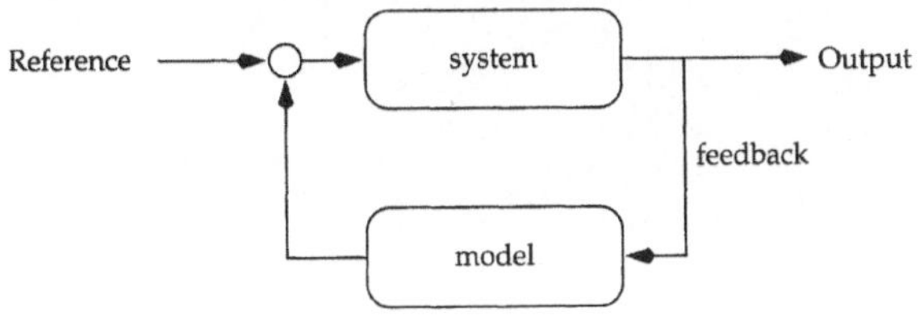


Figure 3.6
Closed-loop control.

The most common example of a control model is the PID (proportional-integral-derivative) controller. This takes the output of the system and compares it with the desired state of the system. It adjusts the control variable $u$ based on the difference between the two values (the error $e$) according to the equation

$$u = Ae + B \int e dt + C \frac{de}{dt}, \tag{3.68}$$

where $A$, $B$, and $C$ are appropriate constants.

The major drawback of this system is that it assumes that the system being modeled is linear or at least behaves in some fashion that is a monotonic function. As the complexity of the system increases it becomes more difficult to formulate that mathematical model.

Fuzzy control plays, in Figure 3.6, the role of the mathematical model and replaces it with another that is built from a number of smaller rules that in general only describe a small section of the whole system. The process of inference binds them together to produce the desired outputs [54]. The inputs and outputs of the system remain unchanged.

The Sendai subway system (Figure 3.7) is the first and most remarkable fuzzy control success story. This system challenged the belief that fuzzy logic could not be used in a safety-driven situation. The Sendai subway system has been in operation since 1986 and is more accurate, has doubled the comfort index, and reduced power consumption by 10% with respect to the previous state of affairs.

Figure 3.7 A train of the Sendai Subway.

### 3.8.2 Nonlinear Simulation and Modeling

Fuzzy techniques can help in describing and simulating systems that cannot be simulated with conventional crisp or algorithmic approaches but can be simulated because of the presence of other information. This is often the case with systems characterized by complex nonlinear dynamics where some of the underlying physical variables might be hard to observe.

Complexity in the system may arise from many factors, such as high dimensionality, too many interacting variables, and unmodeled dynamics such as nonlinearities, time variations, external noise or disturbance, and system perturbations [227].

Fuzzy rule-based systems have been proposed and successfully used to represent simple and complex physical systems. For this purpose, a fuzzy rule-based system consists of a set of rules that represent the expert's understanding of the behavior of the system, a set of input variables affecting the system and a set of output variables, which describe the evolving state of the system. The input and output variables can be numerical, or they can be symbolic or qualitative observations.

### 3.8.3 Decision Making and Support

Decision support systems are computer programs that are able to assist the user in making decisions in a particular domain, often requiring significant specific expertise. Usually, they support the user by accepting a description of the problem as input and proposing possible solutions in response to it.

Historically, a decision support system making use of a large body of domain knowledge, i.e. facts and procedures gleaned from human experts that have proven useful for solving typical problems in a given domain, has been called an *expert system*.

A classical expert system consists of three parts: an inference engine, a knowledge base, and working memory. The methods for representing knowledge that had the biggest impact on the development of expert systems are rules, semantic networks, and frames. Inference engines typically use some meta-knowledge and some reasoning technique, like forward and backward chaining.

A critical step in the construction of an expert system is knowledge acquisition—the act of gathering information for the knowledge base. It is therefore highly desirable to have methods that treat knowledge in the same terms as it is expressed by human experts, that is qualitative, linguistic, and somehow imprecise.

Sources of imprecision in expert systems are diverse. They range from imprecision in solutions, to imprecision in the questions asked to arrive at the solution, and imprecision in knowledge acquisition. Further imprecision can be introduced into the system via incomplete information, limitation of the knowledge representation language, or through the use of weak implications.

In fuzzy expert systems, domain knowledge usually consists of fuzzy rules and the membership functions defining the linguistic values used by the rules.

One area in which the use of fuzzy expert systems is of great interest is that of health care and medical science. In particular, diagnosis and treatment planning are all about uncertain data and qualitative information and fuzzy logic is the technique of choice to approach them.

An example of a recent fuzzy rule-based expert system in the field of orthodontics is Orthoplanner, which uses fuzzy rules with forward and backward chaining [130] to support dentists in deciding orthodontic treatment plans for cases where fixed orthodontic appliance techniques must be employed.

The commercial launch of Orthoplanner took place in September 1994. Currently, Orthopolanner is in use in a number of practices in the United Kingdom.

It has been shown to provide treatment plans which have the same peer support as those produced by an average National Health Service Consultant Orthodontist with a ten-year postgraduate training and experience [212].

### 3.8.4 Classification

Classification, also termed *clustering*, is about dividing a number of data objects into classes, in such a way that objects in one class resemble each other more than they resemble objects in other classes. Such classes are also called clusters of objects, to stress the idea that they are formed by objects that are close together.

Two popular methods for fuzzy classification are the one using fuzzy equivalence relations and the one known as *fuzzy c-means*, after its crisp analog, hard *c-means.*

Much as crisp equivalence relations divide the universe of discourse into equivalence classes, fuzzy equivalence relations cluster elements into partially overlapping fuzzy equivalence classes, grouping together similar elements.

The hard $c$-means method [28] tries to assign all elements in a sample $\mathbf{X} = \{\mathbf{x}_1, \ldots, \mathbf{x}_n\}$ to exactly one out of $2 \leq c < n$ clusters $A_1, \ldots, A_n$, in such a way as to minimize the objective function

$$z = \sum_{i=1}^{n} \sum_{j=1}^{c} \chi_{A_j}(\mathbf{x}_i) d_{ij}^2, \tag{3.69}$$

where $\chi_A$ is the characteristic function of cluster $A$ and $d_{ij}$ is a Euclidean distance between element $\mathbf{x}_i$ and the geometric mean (the *center*) of cluster $A_j$.

The *fuzzy* $c$-means method is an extension to the hard $c$-means obtained by allowing elements to belong to different clusters with different degrees of membership. The objective function of Equation 3.69 thus becomes

$$z = \sum_{i=1}^{n} \sum_{j=1}^{c} \mu_{A_j}(\mathbf{x}_i) d_{ij}^2. \tag{3.70}$$

Let $\mathbf{v}_j$ be the center of fuzzy cluster $A_j$: its $k$th coordinate is now given by

$$v_{jk} = \frac{\sum_{i=1}^{n} \mu_{A_j}^{m}(\mathbf{x}_i) x_{ik}}{\sum_{i=1}^{n} \mu_{A_j}^{m}(\mathbf{x}_i)},$$

where $m \geq 1$ is a weighting parameter controlling the amount of fuzziness of the classification process.

### 3.8.5 Pattern Recognition

While classification aims at establishing the structure in data, pattern recognition takes new data and attempts to assign them to one of a number of known classes. An example of pattern recognition is, given a printed character (the data), to associate it with the letter of the alphabet (the known class) it represents.

Known patterns are usually represented as class structures, where each class structure is described by a number of features. These features define a *feature space*, wherein data are defined. Patterns are fuzzy subsets of this space, whose membership function might be defined in terms of some similarity measure with prototypes, i.e. typical members of classes.

Data can be points of the feature space if feature values can be observed with precision or they can be fuzzy subsets of the feature space if we assume that some features are expressed in qualitative terms or their values are imprecise.

Three popular and easy approaches exist for fuzzy pattern recognition:

- *nearest neighbor*,
- *nearest center* and
- *weighted approaching degree*.

The nearest neighbor method assigns a new data point to the same class as its nearest neighbors among the points whose classification is already known.

In the nearest center approach, each class has a center, and new data points are assigned to the class whose center is nearest.

While both the nearest neighbor and the nearest center approach assume crisp data points, the weighted approaching degree works with fuzzy data. The *approaching degree* [237] between two fuzzy sets, which gives this approach its name, is a metrics assessing the degree of similarity between two sets, i.e. how much two fuzzy sets are overlapping. A new fuzzy data sample is assigned to the class with which there is maximum approaching degree. Since some feature might be more important or relevant than others for the problem at hand, a weight is associated to each feature while calculating the approaching degree with the known classes.

Chapter 4

# Evolutionary Design of Artificial Neural Networks

## 4.1 Introduction

We saw in Chapter 2 that artificial neural networks are biologically-inspired computational models that have the capability of somehow "learning" or "self-organizing" to accomplish a given task. They are particularly efficient when the nature of the task is ill-defined and the input/output mapping largely unknown. However, many aspects may affect the performance of an ANN on a given problem. Among them, the most important is the structure of the neuron connections i.e., the *topology* of the net, the connection weights, the details of the learning rules and of the neural activation function, and the data sets to be used for learning. There are guidelines for picking or finding reasonable values for all of these network parameters but most are rules of thumb with little theoretical background and without any relationship with each other.

Artificial evolution (see Chapter 1) can be used in conjunction with neural networks and it has the potential for addressing several current network design problems. A bio-inspired *rationale* for this approach comes from the study of nervous systems. Today we know that the brain is as much a product of evolution as it is of development and learning. Its overall structure is determined by the information stored in the genotype (the DNA); during development, which is a mapping from genotype to phenotype, this information gives rise to the actual material structure of the brain. But this process is strictly intertwined with cell and connection modifications determined by learning. Thus, by using artificial evolutionary techniques together with the adaptive capabilities of ANN, we can somehow "imitate" an abstract form of the natural processes that give rise to the mature nervous system. Automatically designing ANNs through artificial evolution has advantages over manual design as the complexity of the ANN increases. The *evolutionary engineering* approach is a more integrated and rational way of designing ANNs since it allows single aspects of the design to be taken into account as well as several interacting aspects at once and does not always require expert knowledge. Thus, artificial evolution is the only known way of designing ANNs that covers all aspects of the design problem as an integrated whole.

Evolutionary algorithms can be applied to neural networks in several ways. The most important are: setting the weights in a fixed topology network, determining network topologies, evolving learning rules, and input feature selection. Several systems allow the simultaneous evolution of the architecture and the weights and even of the node transfer function. This field of research is only ten years old and has not yet been systematized. In the following sections we will describe these applications using the recently published literature. Two good review articles on the subject are [251, 241].

## 4.2 Evolving Weights in a Predefined Network Architecture

Let us consider again feedforward, multilayer neural networks. In Chapter 2, Section 2.5 we saw that the standard algorithm for finding suitable weights in the supervised learning regime is backpropagation. However, backpropagation has a number of problems, one of which is the tendency to get stuck at local optima in weight space. Stochastic training methods could be a good alternative to gradient-based ones for network training. Global optimization heuristics, such as simulated annealing and evolutionary algorithms should be more effective for finding suitable weights in multi-modal, non-convex complex search spaces, since they are less prone to get trapped in local optimum points. Here we will concen-

trate on genetic algorithms, another advantage of which is that they also work for nondifferentiable activation functions such as threshold logic units and for recurrent and arbitrarily interconnected networks, which pose problems for gradient-based techniques. In fact, rather than adapting weights based on local improvement only, GAs evolve weights based on the whole network fitness. Early work in the field has been done by Montana and Davis [149] and by Whitley and Hanson [242].

The GA chromosome representing the network can be a list of weights. Usually, each weight in the list would be represented as a binary string to be decoded into real values between, say, −1 and +1. However, binary encoding has a couple of drawbacks for this application (and for many others as well). As the number of connections increase, and it is not uncommon for ANNs to possess hundreds of them, the length of the string increases to a size that severely slows down the evolutionary process. Besides poor scalability, if higher precision is needed in the weight values, more bits have to be added to their binary representation, slowing the search even further. To relieve this problem, real-coded strings can be used instead to directly represent the weight values of a given network. Real-valued chromosomes have many advantages but require specialized genetic operators some examples of which were presented in Chapter 1 for other non-binary alphabets. Montana and Davis [149] chose this kind of representation. They also needed a convention for the ordering of weights in the chromosome: successive weights in the list were assigned to an individual network from top to bottom and left to right. Figure 4.1 depicts this encoding for a small example network (networks with 126 connection weights were used in [149]).

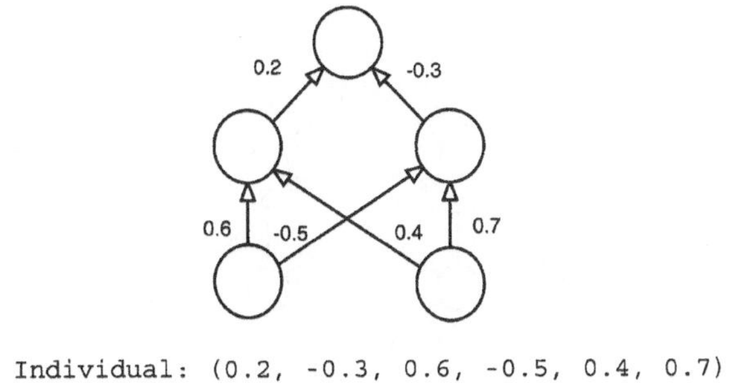


Figure 4.1 GA individual encoding. The weights are represented by a list of real numbers in a predefined order (see text).

After decoding an individual into the corresponding network, its fitness is calculated by computing the cumulated error on the training data (see Chapter 2, Section 2.5). Mutation consists of randomly choosing a non-input neuron and altering its weight by a random value. The recombination operator takes two parent weight vector strings and constructs a single offspring by selecting at random one of the parents' weight for each non-input unit connection from the previous layer. Figure 4.2 illustrates this strategy which is a kind of uniform crossover. Putting the weights of the connections to the same hidden or output node together in the individual representation is beneficial from the point of view of the recombination operator, since otherwise useful evolved feature detectors might systematically be destroyed by crossover. The whole evolutionary process can be described by the following pseudo-code:

*generation* = 0
Assign random weight vectors to the initial population of networks
**while not** *termination condition* **do**
  *generation* = *generation* + 1
  For each genotype, construct the corresponding network
  Compute the fitness of each network on the test data
  Select and reproduce networks according to fitness
  Recombine and mutate selected networks
**end while**

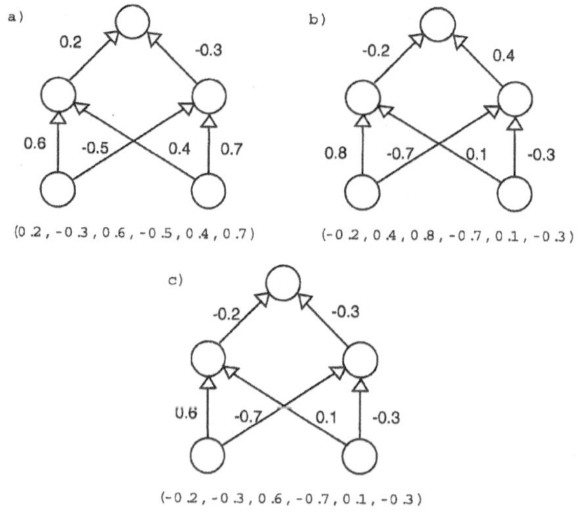


Figure 4.2 Montana and Davis's crossover. Weights for the connections of a single offspring c) are chosen at random from either parent a) or b) (see text).

Figure 4.3 is a schematic illustration of the relationships between the population of weight vector strings and the trained fixed architectures that correspond to this particular encoding. With this technique, Montana and Davis obtained good results on a problem of classification of sonar data as compared to standard backpropagation. This need not always be the case. It has been found that genetic algorithms are rather slow with respect to fast versions of gradient-based methods for supervised applications. When there is gradient information, hybrid methods may be effective (see Chapter 1, Section 1.7.4). Indeed, the efficiency of the evolutionary training can be speeded up by incorporating a local gradient-descent operator. GAs are usually good at locating promising broad regions, while local search may add the needed fine-tuning ability. This process can be seen as a synergetic interaction between evolution and learning [94, 20] and has given better results than using GAs alone on some applications.

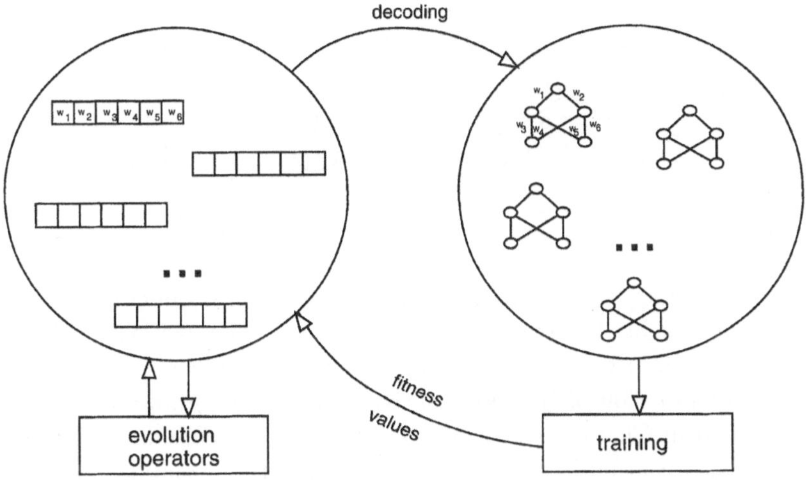


Figure 4.3 Illustration of the artificial evolution of a population of chromosomes encoding the connection weights of a fixed network architecture. The left part of the figure shows the population of strings coding for the weights on which the evolutionary operators are applied. The right part depicts the decoded networks that are subject to fitness evaluation.

Another factor that makes the GA not entirely suitable for training feedforward nets is the so-called *permutation problem* , a topological phenomenon that limits the power of the standard GA crossover operator. The origin of the problem stems from the fact that different linear genotypes may give rise to functionally equivalent networks.

This can be seen for example in Figure 4.4. In general, any permutation of the hidden nodes will produce equivalent networks in terms of network function and, of course, also in terms of the fitness measure. GAs that use standard crossover are likely to create offspring that contain repeated components. That is, because the strings of weights (genotypes) are ordered differently, crossover between such individuals is likely to produce an offspring that contains multiple copies of the same hidden node and will omit other hidden nodes. Highly fit children are thus difficult to result, since the offspring network will be less able computationally than either parent because useful feature detectors associated to hidden nodes during evolution will be lost.

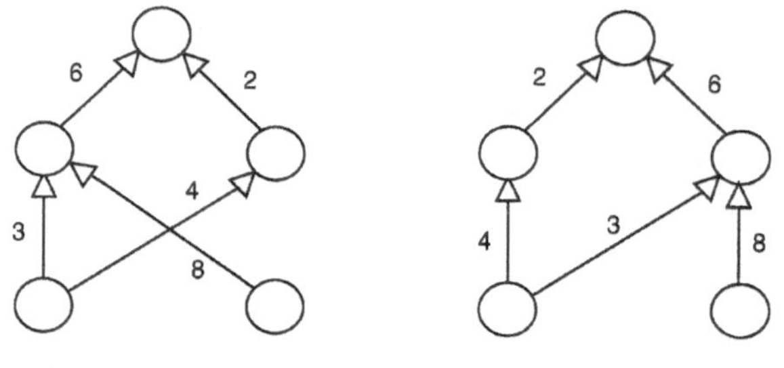


Figure 4.4 Illustration of the permutation problem. The networks are topologically equivalent but their encodings differ.

Since crossover seems to be at the root of several difficulties when evolving weights for fixed structure networks, some researchers have proposed the use of other evolutionary algorithms that do not rely on recombination of individuals. *Evolutionary programming* (EP), described in Chapter 1 Section 1.6.2, is such a technique. The sole source of variation in EP is a sophisticated form of mutation which, by its local character, ensures that individual modifications will not be disruptive. By avoiding crossover, feature detectors are conserved and each network keeps its individual functionality with smooth variation. Fogel *et al.* [69] applied EP to the evolution of connection weights by training a population of networks on a set of standard problems and several successive studies demonstrated the viability of the approach. In all these works the mutation operator randomly alters the weigths according to a Gaussian distribution with variance proportional to some measure of the network error on the task. *Evolution strategies*, being similar to evolutionary programming in their emphasis on advanced mutation operators, are likewise useful for the weight search problem.

### 4.2.1 Genetic Algorithms and Reinforcement Learning Networks

While the usefulness of evolutionary algorithms is questionable for setting weights in cases where established and fast methods already exist, they can be effective when gradient-based methods are not appropriate or cannot be applied. This is the case for *reinforcement learning*, a machine learning technique that was very briefly dealt with in Chapter 2, Section 2.4. To understand the principles, consider an ideal neural network-based robot that can learn. It has sensors to perceive the state of its environments and it can perform a set of actions through its neurocontrol system, such as avoiding obstacles, picking up objects, recharging its batteries, and so on. Such a sensorimotor agent may learn to control its behavior by experimenting in its environment. The robot performs certain actions as a function of the sensory information it gathers along the way. The learning system gives a reinforcement signal, either a positive reward, if the action is considered beneficial, or a punishement if the action decreases the fitness of the agent with respect to the environment. The basic principle is that a GA evolves the synaptic weights such that a positive response causes strengthening of active connections, while a negative response weakens them. Genetic algorithms can be used here even in the absence of precise target output values because they can rely on a relative performance measure for each set of weights. Genetic algorithms have been successfully applied in finding good weights for neurocontrol problems such as the inverted pendulum or some autonomous robot learning tasks [66]. The case study at the end of this chapter deals with the evolution of autonomous adaptive robots in detail, and the reader is referred to that section for an in-depth discussion. Reinforcement learning is described, for example, in an article by Barto [19]. A discussion of evolutionary algorithms in the context of reinforcement learning networks can be found in [241].

## 4.3 Evolving Network Architectures

SELECTION of a suitable network structure is a very important step towards the success of an ANN approach to a given task. If there are too many degrees of freedom, overfitting and poor generalization may result. Thus, a compromise must be found between providing sufficient degrees of freedom for the problem to be well represented and the generalization ability on the task. Choosing a suitable topology implies determining the appropriate number of nodes and their interconnection patterns. Practitioners usually design the network architecture in an unsystematic manner by guesswork and trial and error. But it is true that there are also more refined constructive or pruning heuristics (see Chapter 2, Section 2.5.3, and references therein). Overall, these methods approach the network construction problem in a constrained manner: the space of possible architectures, which is enormous and structurally complex, is searched in a very limited way, slightly changing a given architecture through predefined structural modifications. For instance, most of these methods are limited to a specific class of nets, such as feedforward or recurrent.

Clearly, evolutionary computation with its global search capabilities of multimodal, discontinuous spaces is a promising methodology for the architecture induction problem. There are two major ways in which EAs have been used for searching network topologies: either all aspects of a network architecture are encoded into an individual or a compressed description of the network is evolved. In the first case we speak of *direct* encoding schemes, while the latter leads to so-called *grammatical*, *morphogenetical*, or simply *indirect* encodings.

### 4.3.1 Direct Encoding

In this scheme, the connection topology is represented by means of an *adjacency matrix* that is, an $N$-node architecture is represented by an $N \times N$ matrix $A$, where $a_{ij} = 1$ means that there is a connection between units $i$ and $j$ and $a_{ij} = 0$ stands for no connection. An individual in a population of architectures is simply the string resulting from the concatenation of successive rows of the matrix. This encoding is depicted in Figure 4.5. Actually, the entries $a_{ij}$ may also represent connection weights, in which case both the architecture and the weights can be evolved at the same time. There are several examples of work done with this kind of representation, see for instance references [145, 244]. This encoding is easy to understand and to implement. For example, setting architectural constraints on the network types that will be evolved, such as strict feedforward nets or arbitrarily connected recurrent ones is straightforward.

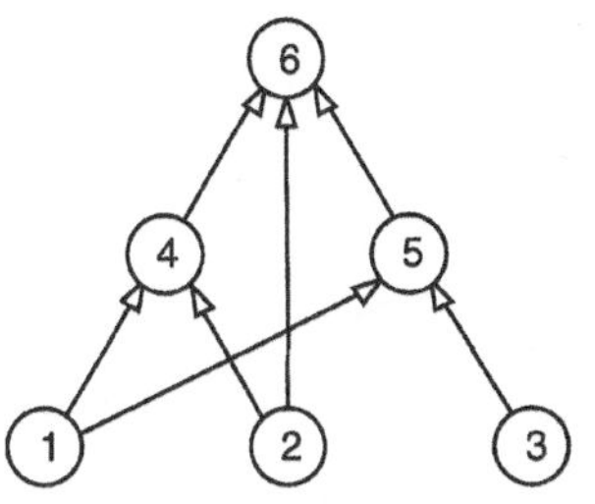


| | 1 | 2 | 3 | 4 | 5 | 6 |
|---|---|---|---|---|---|---|
| 1 | 0 | 0 | 0 | 1 | 1 | 0 |
| 2 | 0 | 0 | 0 | 1 | 0 | 1 |
| 3 | 0 | 0 | 0 | 0 | 1 | 0 |
| 4 | 0 | 0 | 0 | 0 | 0 | 1 |
| 5 | 0 | 0 | 0 | 0 | 0 | 1 |
| 6 | 0 | 0 | 0 | 0 | 0 | 0 |

Figure 4.5 A direct encoding representation scheme. The left part of the figure represents the network architecture, while the right part of the figure shows the corresponding connectivity or adjacency matrix. A 1 at the crosspoint between row $i$ and column $j$ means that there is a connection going from unit $i$ to unit $j$. A 0 entry stands for no connection between the corresponding units.

The evolutionary algorithm works according to the following pseudocode:

*generation* = 0
Initialise the population of individuals
**while not** *termination condition* **do**
  *generation* = *generation* + 1
  For each individual, decode its representation into
  the corresponding architecture
  Compute the fitness of each architecture by training
  it on the test data
  Select and reproduce networks according to fitness
  Recombine and mutate selected networks
**end while**

Figure 4.6 schematically depicts the relationships between genetic evolution of the population of genotypes and the decoding and training process of the actual networks (phenotypes).

The genetic operators can be implemented in several ways. For example, Miller *et al.* [145] used low probability bit mutation and crossover was done by swapping a randomly chosen row between the two selected parents. Once decoded into a network, each individual's fitness is simply calculated by some variant of backpropagation learning as usual. However, it is sometimes useful to include network complexity measures such as the number of nodes and connections

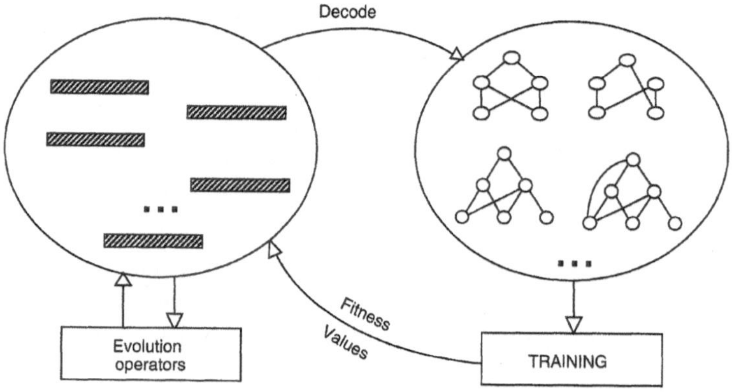


Figure 4.6 In the direct encoding scheme, architectures are fully described by their chromosome string representation. Decoded architectures are trained by some form of supervised learning in order to attribute a fitness value to each one of them. The evolutionary algorithms selects and evolve encoded architectures based on these fitness values.

into the fitness, in order to create selective pressure towards smaller networks. Training time is another fitness criterion that has been often used. We mention here in passing that there is practically no limitation as to how the network's fitness should be defined since continuity and differentiability is not required.

But the direct encoding scheme also has a major drawback: it does not scale well since an $N$-node network potentially has on the order of $N^2$ connections, leading to very long chromosomes. Although the size of the matrices can be reduced by using constraints coming from previous knowledge on the architectures to be evolved, the scaling properties of the representation are poor. The permutation problem that was described in the previous section is still there and, moreover, training a whole population of networks by backpropagation or similar methods can be extremely slow. Direct encoding is thus only useful for small architectures.

As we noted at the beginning of the section, there is the possibility of evolving the architecture and the weights of the network at the same time [31, 4]. In this context, the work of Angeline *et al.* [4] is interesting because it avoids some of the problems related to recombination and can evolve a wide spectrum of ANN architectures.

In [4] Evolutionary Programming (EP) was used. In EP, the representation evaluated by the fitness function is directly submitted to a single evolutionary operator: mutation. A form of *structural* mutation is used for architecture evolution and, at the same time, the connection weights are also altered by *parametric* mutations. Parametric mutations are similar to those described in the previous section on the evolution of weights in a fixed architecture [69]. Structural mutations alter the number of hidden nodes and the connectivity between all nodes, with some restrictions concerning input and output nodes. Structural mutations attempt to preserve the behavior of a network by making changes as smooth as possible. This is different from genetic algorithms where recombination may strongly disrupt the continuity between parent and child, often destroying useful building blocks. The methodology has been demonstrated on a number of test cases with good results and little restriction on the architecture typology.

Another effort in this direction is the recent work of Yao and Liu [253] in which the evolutionary system called EP-Net is proposed. EP-Net is based on evolutionary programming with several different sophisticated mutation operators. The operators are designed in such a way as to maintain close links between parents and their offspring. EP-Net tries to evolve the behavior of the ANN and simultaneously determines the architecture and the weights, including biases. Although an explicit measure of parsimony is not used in EP-Net, the system favors simpler and more effective networks by preferring node or connection deletion rather than node or connection addition, if the performance is not lowered by the move. The system has been tested on a number of standard benchmark problems with good results.

The same authors recently proposed making explicit use of population information [254]. Instead of just picking the best ANN in the last generation as the final result, Yao and Liu suggest forming the result by combining in a certain way the individuals in the last generation in order to exploit all the information contained in the whole population. The idea is that a population contains more information than any isolated individual and that such information can be suitably used to improve generalization of the learning system. The appproach has been tested on real world data sets, showing that there are simple linear combination methods of individuals in the last generation that always produce results outperforming the best individual.

Genetic programming has been used by Koza as a flexible system for finding both the architecture and the weights of a neural network [117]. Koza and coworkers used trees as building blocks for representing neural networks. Although they were able to evolve neural networks for solving a few standard tasks, the tree representation appeared not to be flexible enough for evolving general purpose ANNs. In the next section we will see other more useful ways of using genetic programming-like techniques for discovering neural network architectures and weights for a given task.

Most of the applications of evolutionary algorithms to date have been made on multilayer feedforward or recurrent architectures. A notable exception is the work of Polani and Uthmann [176] who applied a GA to find improved topologies for Kohonen's feature maps (Chapter 2, Section 2.6.3). In this study, GAs were applied to create self-organizing maps able to adapt to a given input space without imposing a predefined topology. It was found that supposedly optimal "flat" topologies do not always give the best results. If these findings are confirmed, the technique could thus be useful for constructing network topologies for Kohonen's maps with better convergence speed and adaptation properties.

### 4.3.2 Indirect Encoding

In view of the scalability problems brought about by direct encoding methods and of their consequences in terms of performance, several researchers focused their efforts on techniques for developing or *growing* neural networks, rather than looking for a complete network description at the individual level. In general terms, these so-called *grammatical* encodings are variations or extensions of Lindenmayer's L-system [127, 181], a string-rewriting system which was originally introduced as a mathematical model of plant development. The purported advantages of grammatical encodings are better scalability and the possibility of finding building blocks of general utility and of reusing developmental rules for general classes of problems. Moreover, the destructive effect of crossover should be less relevant for developmental rules than for direct encodings. In a way, this approach is the closest in spirit to the growth and development of biological neural networks, although at a very much simpler and abstract level.

The early work of Kitano [112] is representative of this methodology and will be described in some detail. In Kitano's methodology, a graph-generation grammar is encoded into a chromosome. A standard GA is used to evolve a population of such network-generating grammars, fitness being measured after "morphological development", whereby an actual ANN architecture is generated from its grammar description.

A grammar consists of one or more *productions*. A production is a rewriting rule that associates a left-hand side called a *head* to a right-hand side called a *body*. The two sides are separated by the metasymbol $\rightarrow$. The symbol $\rightarrow$ indicates that the category to the left of it can be composed of the category to the right; that is, it "can be rewritten as" the category to the right. For example, the following productions define a *number* grammatically:

$$\langle digit \rangle \rightarrow 0 \mid 1 \mid 2 \mid 3 \mid 4 \mid 5 \mid 7 \mid 8 \mid 9$$
$$\langle number \rangle \rightarrow \langle digit \rangle$$
$$\langle number \rangle \rightarrow \langle number \rangle \langle digit \rangle$$

The symbols $0 \ldots 9$ are called *terminals*. Terminals cannot appear on the left-hand side of a production. The symbol $\mid$ stands for "or". That is, when applying the production, only one of the terminals can be chosen. To construct a $\langle number \rangle$, let us start with the third production and replace 2 for $\langle digit \rangle$: this gives us $\langle number \rangle \rightarrow \langle number \rangle 2$. Now rewrite $\langle number \rangle$ on the right using again the same production with 4 as $\langle digit \rangle$: $\langle number \rangle \rightarrow \langle number \rangle 42$, finally use production two with $\langle digit \rangle = 7$ in place of $\langle number \rangle$ on the right: $\langle number \rangle \rightarrow 742$. That is, the numeric string 742 has been generated. Since all the symbols on the right are terminals, the generation is complete.

Kitano applied the above idea to generate connection matrices for networks by means of a set of graph generation rules. Figure 4.7 shows an example of a graph-generation grammar. Each production rule consists of a left-hand side which is a non-terminal and a right-hand side which is a $2 \times 2$ matrix. The right-hand side can be a terminal or a non-terminal; in the latter case the matrices are binary i.e., they have elements that can only be 0 or 1. By successively applying the productions from a given start symbol one ends up with a binary array which can be interpreted as the adjacency matrix of a directed graph representing a neural network. Since each

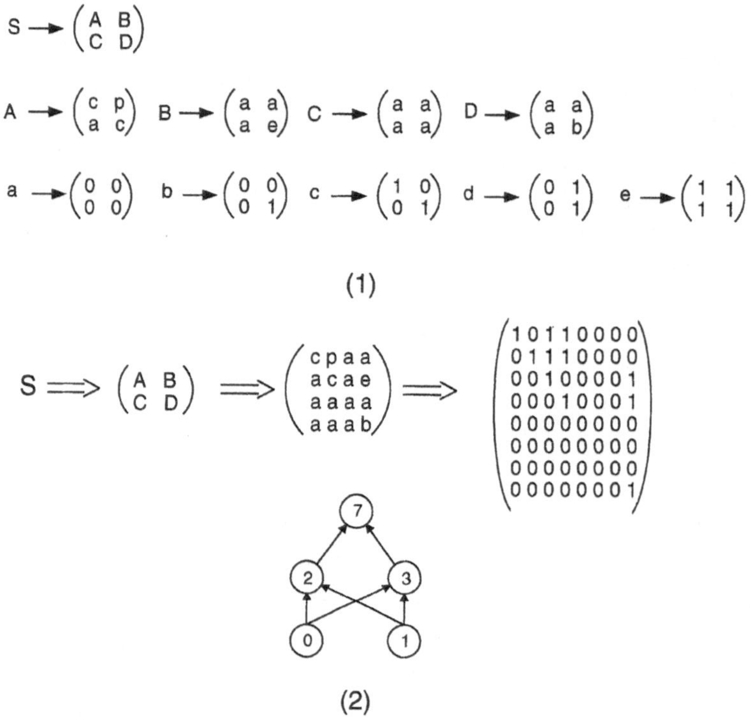

$$S \rightarrow \begin{pmatrix} A & B \\ C & D \end{pmatrix}$$

$$A \rightarrow \begin{pmatrix} c & p \\ a & c \end{pmatrix} \quad B \rightarrow \begin{pmatrix} a & a \\ a & e \end{pmatrix} \quad C \rightarrow \begin{pmatrix} a & a \\ a & a \end{pmatrix} \quad D \rightarrow \begin{pmatrix} a & a \\ a & b \end{pmatrix}$$

$$a \rightarrow \begin{pmatrix} 0 & 0 \\ 0 & 0 \end{pmatrix} \quad b \rightarrow \begin{pmatrix} 0 & 0 \\ 0 & 1 \end{pmatrix} \quad c \rightarrow \begin{pmatrix} 1 & 0 \\ 0 & 1 \end{pmatrix} \quad d \rightarrow \begin{pmatrix} 0 & 1 \\ 0 & 1 \end{pmatrix} \quad e \rightarrow \begin{pmatrix} 1 & 1 \\ 1 & 1 \end{pmatrix}$$

(1)

$$S \Longrightarrow \begin{pmatrix} A & B \\ C & D \end{pmatrix} \Longrightarrow \begin{pmatrix} c\,p\,a\,a \\ a\,c\,a\,e \\ a\,a\,a\,a \\ a\,a\,a\,b \end{pmatrix} \Longrightarrow \begin{pmatrix} 1\,0\,1\,1\,0\,0\,0\,0 \\ 0\,1\,1\,1\,0\,0\,0\,0 \\ 0\,0\,1\,0\,0\,0\,0\,1 \\ 0\,0\,0\,1\,0\,0\,0\,1 \\ 0\,0\,0\,0\,0\,0\,0\,0 \\ 0\,0\,0\,0\,0\,0\,0\,0 \\ 0\,0\,0\,0\,0\,0\,0\,0 \\ 0\,0\,0\,0\,0\,0\,0\,1 \end{pmatrix}$$

(2)

Figure 4.7 Example of the use of graph-generation production rules for the development of a feedforward exclusive-or network. The grammar is depicted in the upper part (1). The lower part (2) of the figure shows the successive application of the grammar rules starting from symbol $S$ and producing a network for the computation of the exclusive-or Boolean function. The figure is adapted from Kitano's article [112].

non-terminal symbol in the grammar has only one right-hand side, the generated network is unique. The example shown in the Figure depicts the development of a neural network which is able to compute the exclusive-or function if suitable weights are provided.

A GA was used for evolving a population of individuals consisting of a grammar representation of networks. Each individual has a number of separate components which correspond to the production rules of the grammar. Each rule is represented in five positions. The first rule begins with the start symbol $S$ and is followed by four symbols chosen from the set $A - Z$. This is always required to get the development process started. All the other rules have a left-hand side of a rule in the first position, followed by four symbols chosen from the set $a - p$. The rules that transform a symbol in $a - p$ to a binary $2 \times 2$ matrix are fixed and do not appear in the individual representation. Figure 4.7 graphically depicts the structure of the chromosome. The GA used in [112] was a rather standard one, with fitness proportionate reproduction, elitism, variable mutation rate, and single and multi-point crossover. The fitness of an individual is evaluated by first building a network from the grammar encoded in the individual and then training the network by backpropagation on a given task. As in other studies, the sum of the square of the errors was used as the quality criterion. As Yao and Liu [253] indicated, this method of evaluating fitness is very noisy and inaccurate because the fitness obtained depends on the random initial weights.

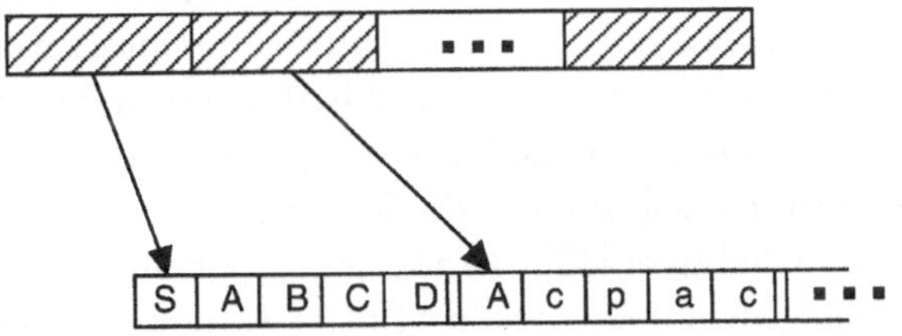


Figure 4.8 Schematic illustration of a graph-generation grammar encoded into a GA individual. The figure is adapted from Kitano's article [112].

Kitano applied his grammatical approach to various sizes of encoder/decoder problems, comparing the results with those obtained from direct encoding methods. He concluded that the grammar encoding method converges much faster, generates more regular network structures, and scales better with respect to the network size. But Kitano's test cases were very simple and, on the other hand, recent work by Siddigi and Lucas [209] has shown that Kitano's encoding scheme does not provide an advantage over the direct encoding scheme. Although no general conclusions as to whether the approach is actually superior can be drawn yet, grammatical encoding is an interesting idea that has been further pursued by other researchers. F. Gruau, in particular, developed a grammar based developmental technique called *cellular encoding* [81] in which genetic programming is used to evolve the architecture of neural networks. Cellular encoding is a powerful but rather complicated method that cannot be reasonably explained succinctly. Reference [241] has a readable introductory description of the technique. Using cellular encoding, Gruau and coworkers have been able to grow artificial neural network architectures and weights for various difficult tasks, including recurrent network architectures (see [241] and references therein). The weights could be at first only +1 or −1, that is only Boolean networks could be developed. Later, real weighting schemes were included in the evolutionary process, as well as biases and thresholds and a form of automatic function definition that allows reuse of subtrees and results in more modular architectures.

## 4.4 Evolution of Learning Rules

AN INTERESTING application of evolutionary algorithms to the design of neural networks is the discovery of learning rules for adjusting the connection weights. In Chapter 2, Section 2.4 we saw that there are several standard learning rules. However, when there is little knowledge about the most suitable architecture for a given problem, the automatic and possibly dynamic adaptation of the learning rule becomes very useful. Evolutionary algorithms are eminently suited to the task. One of the first studies in this direction was conducted by Chalmers [42]. In this work, Chalmers's aim was to see if the well-known delta rule, or a fitter variant, could be evolved automatically by a GA. A number of assumptions were made at the outset in order to restrict the form of the learning algorithm to linear functions of the relevant parameters i.e., the input, output, and target values, as well as the weight change and a scale parameter similar to the learning rate constant. The pairwise products of the parameters were also used in the linear combination. The network architecture was feedforward with input and output layers only, which can only learn mappings that are linearly separable. With a suitable chromosome encoding and using a number of linearly separable mappings for training, Chalmers was able to evolve a rule analogous to the delta rule, as well as some of its variants. Although this study was limited to somewhat constrained network and paramcter spaces, it paved the way for further progress. It is still unknown wheter artificial evolution can discover efficient learning algorithms for complex networks. However, it is apparent that genetic programming should prove particularly suitable for the task of discovering new rule forms and, of course, coevolution of architectures, weights, and learning rules could take place simultaneously, although the search space would be enormous in this case. Reference [251] presents a summary of the work done in the field in the last few years.

## 4.5 ANN Input Data Selection

STATISTICAL methods, such as principal components analysis, are customarily used to select input features to a neural network in order to reduce or combine the data into a statistically significant and effective sample. Indeed, in many works using neural networks, no particular effort is made to select appropriate input features, which obviously detracts from efficiency by slowing down training and may lead to poor generalization capabilities. Evolutionary algorithms are an attractive alternative to statistical methods for dimensionality reduction of the input data set. Several studies have shown that searching the space of input data for optimal or nearly optimal compact subspaces can be done effectively with evolutionary algorithms, substantially reducing the size of the input data without a loss in performance. Essentially, in an EA encoding of the problem, each individual in the population represents a portion of the input data. The ANN is trained with these vectors and the result is part of the individual's fitness. Two references in the field are [43, 32].

A related issue is the partitioning of the input data of a network into a training and a validation set. This partitioning is almost always done quite arbitrarily, although it may influence the network performance in a significant manner. Reeves and Taylor [186] recently applied genetic algorithms to select training sets for radial basis function neural networks. They found better generalization performance for an artificial benchmark problem and for a real-world classification problem over randomly selected training data. Since a fixed architecture was used here as well as in [43, 32], this raises the issue of the mutual influence between the architecture and the training data. When the correct architecture is not known, it may well be that different architectures would work best with different training sets on a given problem. The possibility of co-evolution of training sets and network architectures has been investigated by Mayer [138] by using a symbiotic approach in which independent populations of ANNs and traning sets are co-evolved by a GA. The fitness of an ANN is shared between the net and the data set it has been trained with. The conclusions were not clear-cut but the idea, which is an extension of previous works by Hillis and Paredis [93, 164], is interesting.

## 4.6 Evolution of Neural Machines

ARTIFICIAL neural networks are usually simulated on a conventional general-purpose computer. However, they actually represent attractive alternative computational models and it would make sense to develop dedicated hardware for neurocomputing, especially for on-board and special-purpose applications. In fact, if executed on an appropriate hardware, these machines would be much faster than the corresponding simulations and they would reach their real potential, being closer architecturally to the inspiring biological systems and enjoying the far superior circuit speeds of modern electronics or other technologies. A number of neural chips have been designed and built to date. However, quite often special-purpose machines are the victims of technological advances, in the sense that today's fast specialized hardware will probably be slow compared to tomorrow's simulations on state-of-the-art general-purpose hardware. Nevertheless, dedicated machines are often required in many applications and neuro-hardware has been implemented using different technologies: digital and analog circuits, optical architectures and, more recently, even biological and chemical chips.

A short review of the subject will be presented in Chapter 8. However, in the context of evolutionary neural networks, it is worth noting that promising hardware technologies exist that do not require the system to be designed and wired-up once and for all, thus allowing for real-world applications that may change in real-time. This new research domain called *evolvable hardware* which is less than ten years old, is based on the idea of using a particular kind of circuit whose architecture and functions may evolve dynamically and autonomously as a function of task requirement. Thanks to the existence of reconfigurable hardware devices such as field-programmable gate arrays (FPGA) this seemingly odd idea can be exploited in practice. An FPGA is an array of logic cells (see Figure 4.9) whose functions and interconnections may be programmed by means of a string of *configuration bits*. FPGAs can be reconfigured at will and very quickly by just downloading a new configuration string into the chip.

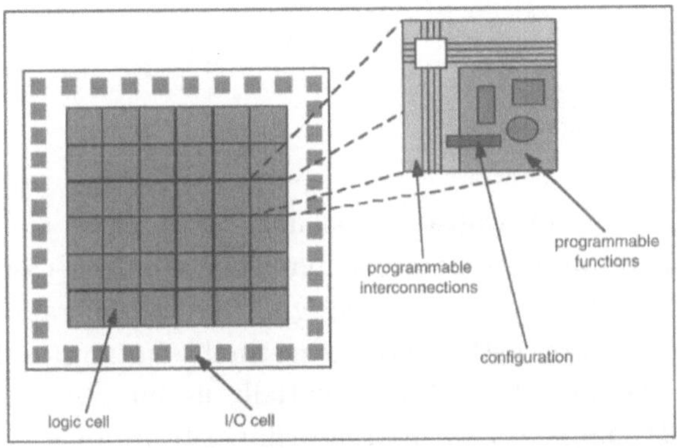


Figure 4.9 Schematic representation of an FPGA.

The key idea of evolvable hardware is to consider the bit string as a genetic algorithm chromosome. The genetic algorithm evolves a population of bit strings such that, in time, better individuals—i.e. bit strings—are likely to emerge. Given a properly designed fitness function, suitable hardware structures for a given task can then be evolved by evaluating each individual chromosome of a population through configuring and measuring the adequacy of the corresponding circuit to the task. This straightforward concept is illustrated in Figure 4.10. There is a number of subtle points lurking behind the scenes for the realization of this neat idea. First of all, configured devices should not crash, or damage their measuring environment when meaningless bit configurations are sent to them. Moreover, reconfiguration and fitness measure should be very fast, in order not to slow down the evolutionary process. Modern FPGAs, such as the Xilinx 6200 series, do indeed possess these features to some extent and can be used, although existing reconfigurable devices were not designed with evolution in mind from the start.

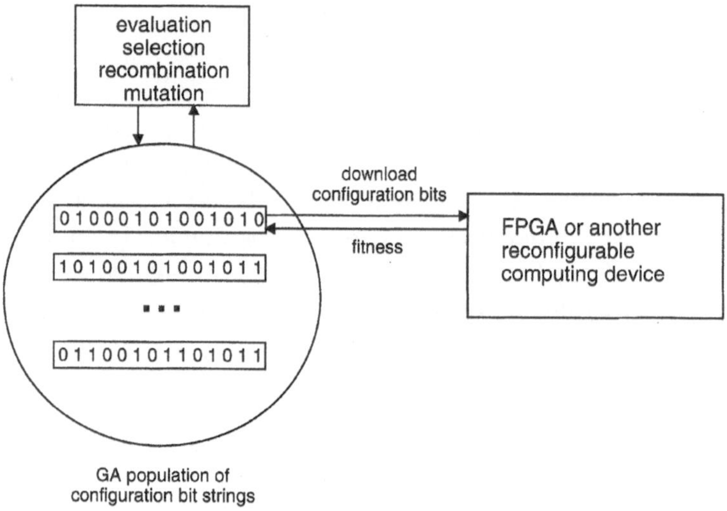


Figure 4.10
Illustration of the evolution of configuration bit strings by a GA. Each individual is downloaded into an FPGA where the fitness of the resulting circuit is evaluated and sent back to the GA.

When speaking of evolvable machines, it is useful to distinguish between *offline* and *online* evolution. The simplest case of offline evolution is *evolutionary circuit design*, where all operations are carried out in software, with the resulting final solution possibly loaded onto a real circuit. Though a potentially useful design methodology, this falls completely within the realm of traditional evolutionary techniques.

In the case depicted in Figure 4.10, although a real circuit is used during the evolutionary process, we still speak of offline hardware evolution since most operations are carried out offline in software. In effect, the population is stored in an external computer, which also controls the evolutionary process. Ideally, in online evolution one would have all operations (selection, crossover, mutation), as well as fitness evaluation carried out in hardware. Building machines that would evolve and even adapt autonomously during their lifetime is an enticing perspective, but all practical systems to date have been evolved using the scheme of Figure 4.10.

Evolvable hardware has been used to automatically design a number of computing machines including cellular automata, data compression and decompression devices, analog circuits and autonomous robot controllers. The interested reader can find articles covering basic issues and current research in the field in references [11, 135, 252]. Here we are interested in neural networks hardware evolution, two prominent examples of which being the works of Higuchi *et al.* [153] and of de Garis *et al.* [53]. We mention in passing that both de

Garis and Higuchi played a leading role in the establishment of the foundations of the field in the early 1990s.

### 4.6.1 Evolvable Neural-Network Hardware

Higuchi *et al.* observe that most industrial applications of artificial neural networks are limited to systems in which learning takes place offline, before the network is actually used for recognition tasks on unseen data. This obviously prevents the system from making real-time adjustments according to changing requirements and therefore causes a lack of online adaptation capabilities. We have seen that evolutionary methods are useful for the overall ANN design process. However, until now these methodologies have been mostly limited to computer simulations of the actual nets. Genetic algorithms and reconfigurable hardware are the key ingredients for truly adaptable and fast networks. Thus, dynamically reconfigurable ANNs possess the potential to be useful for embedded applications where compact, adaptable, and fast machines are preferable to either simulations or specialized ANN hardware.

The GRD (Genetic Reconfiguration of Digital signal processors) chip [153] is an evolvable hardware system for neural network applications designed with optimality, adaptability, and efficiency issues in mind. The system consists of a 32-bit RISC processor and a binary tree of 15 DSPs (Digital Signal Processors). The GA runs on the RISC processor thus avoiding the need for a host machine for this task. With rapidly declining hardware costs, it will be common in the future to have a lot of processing power available, even for special-purpose applications. Both the topology and the hidden layer node transfer functions of the neural net can be dynamically reconfigured using a GA. Figure 4.11 schematically illustrates how the configuration and learning take place. A GA individual (or chromosome) represents a particular network architecture, including the choice of the transfer function which can be either a Gaussian or a signoid. The upper part of the figure shows the architecture of two evolved networks. These networks are mapped onto the GRD chips which are programmable function units, that is, their function can be changed in real time by just downloading another configuration string into them, a fact that is illustrated in the lower portion of the figure. New networks are obtained through the GA by applying the

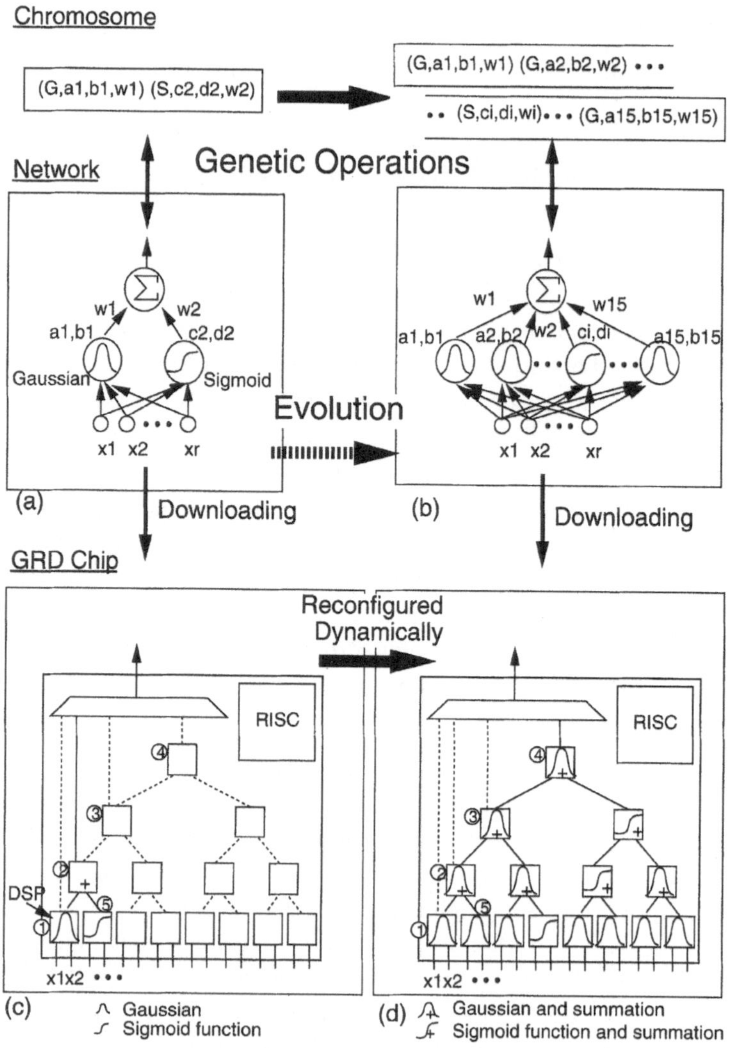


Figure 4.11
Illustration of hardware evolution of the GRD chip for neural network online applications. Figure reproduced by permission of the authors [153].

usual genetic operators. In the figure, the network on the left with three nodes is finally evolved into the 15-node network on the right. The fitness of a network is calculated as the sum of the squared error over the training set by using local learning with steepest descent. The binary tree architecture allows the parallel calculation of the sum of node activations and is thus useful for applications that require high performance. Alternatively, a single DSP can be used by time slicing.

The GRD chip has been employed in test applications of digital mobile communications with excellent results. The planned use includes a number of applications whose environments vary over time and have real-time constraints.

### 4.6.2 Evolving Digital Brains

Standard ANN systems do not comprise a very large number of neurons: a few tens to a few hundreds of them being the rule. On one hand, statistical and computational properties are more difficult to analyse for large networks, unless there is a lot of regularity in the interconnections, such as in cellular neural networks. On the other hand, if the system is to be simulated on a conventional computer, which is often the case, a large number of units may slow down learning to unacceptable levels. Small artificial neural networks have proved very useful in many fields and today they are standard components of common devices and appliances in industry and the consumer market. However, if one could afford to directly build adaptive neural machines in hardware, then it would be worthwhile to self-assembly a large number of computing units by evolution and learning to see if some higher-level properties could emerge that are a better match to the biological systems they imitate. Of course, with millions, or even billions, of self-assembling units, we will be forced to abandon to some extent the idea of being able to analyze how the machine works in detail, a prospect which can be upsetting for some people. This is not necessarily as frightening as it sounds: after all, nobody really understands how the brain works in detail, nevertheless we all make use of it daily without ever noticing. Statistical physics is another example of a perfectly valid description of natural systems in which only average quantities over zillions of atoms and molecules can actually be observed, the fine behavior of a single atom being totally irrelevant. de Garis's CAM-Brain project, an ambitious and visionary research endeavor, is structured along these lines. In de Garis's own words:

> The original (perhaps over-ambitious) aim of our "CAM-Brain Project", as stated at its beginning in 1993, was to build an artificial brain with a billion artificial neurons, by the year 2001, using evolved cellular automata (CA) based neural circuit modules. In reality, 6 years later, this number will be maximum 40 million neurons and 32,000 modules. These CA based neural network

> modules grow and evolve at electronic speeds inside special FPGA based hardware called a CAM-Brain Machine (CBM), which can update CA cells at a rate of 150 billion a second, and can evolve a neural net module in about 1 second. This speed should make brain building practical. Tens of thousands and more of these evolved modules can be assembled into humanly defined artificial brain architecures. The evolved CA based circuit modules are downloaded into a large RAM space and updated by the CBM fast enough for real time control of a kitten robot.

The CAM-Brain machine (CAM stands for Cellular Automata Machine) is a research tool for the simulation of huge assemblages of artificial neurons, "artificial brains" in de Garis's language. A small sample of an evolved portion of a neural circuit is depicted in Figure 4.12.

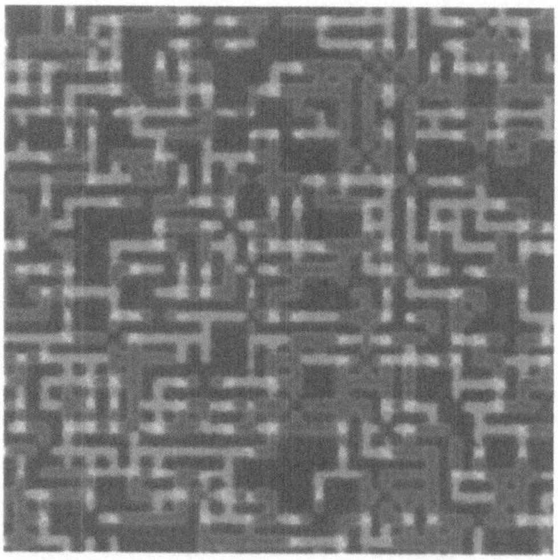

Figure 4.12 Small enlarged region of a CAM-BRAIN 2D 10,000,000 neuron circuit sample. Reproduced by permission of the author.

The CBM machine is specially designed using dynamically reconfigurable FPGA chips to support the growth and signalling of neural networks in two and three dimensions. The originality of the approach lies in the fact that the neural modules are unusually large and they are not designed in the customary way; rather, they are evolved directly in hardware using genetic algorithms, in the manner previously outlined in the section. Figure 4.13 is a photograph of the newly released CBM machine.

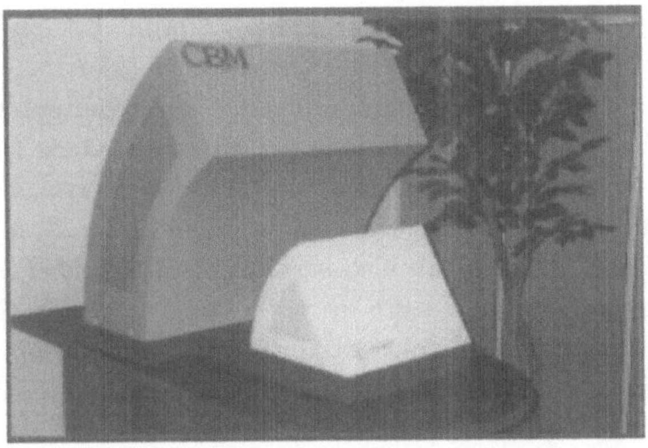


Figure 4.13 The CAM-Brain Machine (CBM) is a piece of specialized "evolvable hardware" for fast evolution of cellular automata-based neural network circuit modules. The shape is supposed to represent a slice of cortex, and its colors (grey and white) represent the outer "grey matter" of the cortex (i.e. the neurons) and its inner "white matter" (the interconnecting axons). Figure reproduced by permission of the author.

This is a large-scale ongoing project. Results on simple learning tasks to date are encouraging (see for instance [53]). The current research aims at controlling the kitten-robot which will be used as a demonstration vehicle to show the capabilities of the evolved artificial brain. The robot will be life-size and its "brain" will be offline, controlling a whole range of sensor and motor behaviors via a radio link.

## 4.7 A Case Study: Evolutionary Autonomous Robots

An autonomous robot is a mechanical device that can operate without being attached to a power supply or an external computer. Ideally, the aim is for the robot to be able to adapt to unpredictable sources of change. This is a toll order and there are as yet few experimental autonomous robots that are able to function correctly in a restricted but changing environment.

Evolutionary robotics is advocated as an automatic method to discover efficient controllers for robots that operate in real environments. The situated nature of the evolutionary approach is such that often evolved controllers find surprisingly simple, yet efficient, solutions that capitalize upon unexpected invariants of the interaction between the robot and its environment. The remarkable simplicity and efficiency of these solutions is a clear advantage for fast and real-time operation required from autonomous robots, but it raises the issue of robustness when environmental conditions change. Environmental changes can also be a problem for other approaches (programming, learning, e.g.) to the extent to which the sources of change have not been considered during system design, but they are even more so for evolved systems because these often rely on environmental aspects that are often not predictable by an external observer.

A useful approach consists of combining evolution and learning "during life" of the individual (see [21] for a comprehensive review of the combination of evolution and learning). This strategy not only can improve the search properties of artificial evolution, but can also make the controller more robust to changes that occur faster than the evolutionary time scale (i.e., changes that occur during the life of an individual) [159]. This is typically achieved by evolving neural controllers that learn with an off-the-shelf algorithm, such as reinforcement learning or back-propagation, starting from synaptic weights specified on the genetic string of the individual [2]. Only initial synaptic weights are evolved. A limitation of this approach is the "Baldwin effect" (see [94] for an example of the Baldwin effect in a computational model), whereby the evolutionary costs associated with learning give a selective advantage to the genetic assimilation of learned properties and consequently reduce the plasticity of the system over time.

In previous work Floreano and Mondada[64] have suggested *evolving the adaptive characteristics* of a controller instead of combining evolution with off-the-shelf algorithms. The method consists of encoding on the genotype a set of four local Hebb rules (see Chapter 2, Section 2.6.1) for each synapse, but *not the synaptic weights*, and let these synapses use these rules to adapt their weights online starting always from random values at the beginning of the life. Since the synaptic weights are not encoded on the genetic string, there cannot be genetic assimilation of abilities developed during life. In other words, these controllers can rely less on genetically-inherited invariants and must develop in real time the connection weights necessary to achieve the task. When comparing evolution of genetically-determined weights with evolution of adaptive controllers on a simple navigation task, it has been shown that the latter approach generates similarly good performances in less generations [65] by taking advantage of the combined search methods and that evolutionary adaptive controllers can adapt to environmental changes that involve new sensory characteristics and new spatial relationships of the environment [226].

A set of experiments to test the robustness of this approach to environmental changes are presented here addressing two important types of change for robot controllers: the transfer of evolved controllers from a simulated robot to a physical robot (Khepera) and across different robots. The results are compared to those obtained from evolution of genetically-determined weights and evolution of noisy synaptic weights (control condition). Evolutionary adaptive controllers not only report significantly better performances, but also display qualitatively different ways of coping with the task at hand. More details can be found in the original article [225].

### 4.7.1 Method

The controller that was used in the experiments is a fully-recurrent discrete-time neural network (Figure 4.14). It has access to three types of sensory information from the robot:

1. *Infrared light*: the active infrared sensors positioned around the robot (Figure 4.15, a) measure the distance from objects. Their values are pooled into four pairs and the average reading of each pair is passed to a corresponding neuron.
2. *Ambient light*: the same sensors are used to measure ambient light too. These readings are pooled into three groups and the average values are passed to the corresponding three light neurons.
3. *Vision*: the vision module (Figure 4.15, b) consists of an array of 64 photoreceptors covering a visual field of 36° (Figure 4.15, center). The visual field is divided up in three sectors and the average value of the photoreceptors (256 gray levels) within each sector is passed to the corresponding vision neuron.

Two motor neurons are used to set the rotation speed of the wheels (Figure 4.15, c). Neurons are updated every 100 ms according to the

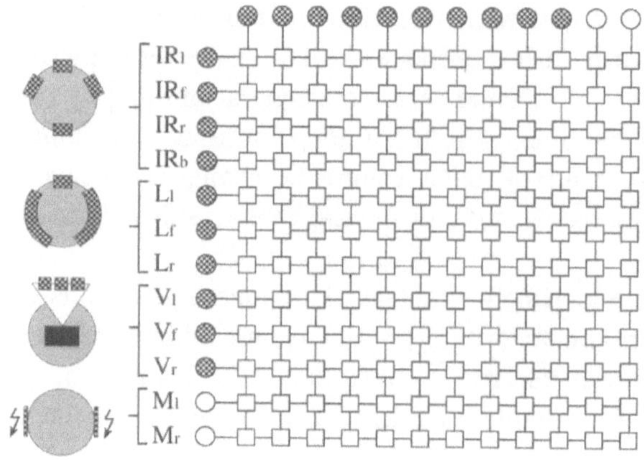


Figure 4.14 The neural controller is a fully-recurrent discrete-time neural network composed of 12 neurons giving a total of 12 x 12= 144 synapses (here represented as small squares of the unfolded network). 10 sensory neurons receive additional input from one corresponding pool of sensors positioned around the body of the robot shown on the left (l=left; r=right; f=front; b=back). $\vec{IR}$=Infrared Proximity sensors; $\vec{L}$=Ambient Light sensors; $\vec{V}$=vision photoreceptors. Two motor neurons $\vec{M}$ do not receive sensory input; their activation sets the speed of the wheels ($M_i > 0.5$ forward rotation; $M_i < 0.5$ backward rotation).

following equation:

$$y_i \leftarrow \sigma\left(\sum_{j=0}^{N} w_{ij} y_j\right) + I_i,$$

where $y_i$ is the activation of the $i$th neuron, $w_{ij}$ is the strength of the synapse between presynaptic neuron $j$ and postsynaptic neuron $i$, $N$ is the number of neurons in the network, $0 \leq I_i < 1$ is the corresponding external sensory input, and $\sigma(x) = (1 + e^x)^{-1}$ is the sigmoidal activation function. $I_i = 0$ for the motor neurons.

Each synaptic weight $w_{ij}$ is randomly initialized at the beginning of the individual's life and can be updated after every sensory-motor cycle (100 ms),

$$w_{ij}^t = w_{ij}^{t-1} + \eta \Delta w_{ij},$$

where $0.0 < \eta < 1.0$ is the learning rate and $\Delta w_{ij}$ is one of the four modification rules specified in the genotype which may co-exist within the same network:

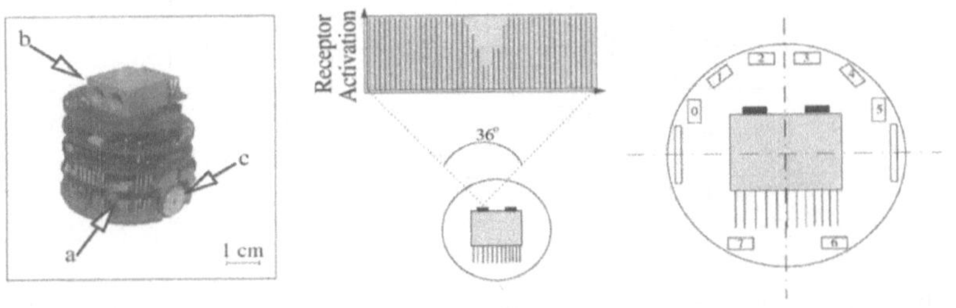


Figure 4.15 The Khepera robot used in the experiments. Infrared sensors (a) measure object proximity and light intensity. The linear vision module (b) is composed of 64 photoreceptors covering a visual field of 36° (center). The output of the controller generates the motor commands (c) for the robot. Right figure shows the sensory disposition of the Khepera robot.

1. *Plain Hebb rule*: strengthens the synapse proportionally to the correlated activity of the two neurons.
2. *Postsynaptic rule*: behaves as the plain Hebb rule, but in addition it weakens the synapse when the postsynaptic node is active but the presynaptic is not.
3. *Presynaptic rule*: weakening occurs when the presynaptic unit is active but the postsynaptic is not.
4. *Covariance rule*: strengthens the synapse whenever the difference between the activations of the two neurons is less than half their maximum activity, otherwise the synapse is weakened.

Synaptic strength is maintained within a range $[0, 1]$ (notice that a synapse cannot change sign) by adding to the modification rules a self-limiting component inversely proportional to the synaptic strength itself.

Two types of genetic (binary) encoding are considered (see table):

1. *Synapse Encoding*: also known as direct encoding [251], every synapse is individually coded on five bits, the first bit representing its sign and the remaining four bits its properties (either the weight strength or its adaptive rule).
2. *Node Encoding*: each node is characterized by five bits, the first bit representing its sign and the remaining four bits the

| Encoding | Bits for one synapse / node | | | | |
|---|---|---|---|---|---|
| Genotype | 1 | 2 | 3 | 4 | 5 |
| A | sign | strength | | | |
| B | sign | Hebb rule | | rate | |
| C | sign | strength | | noise | |

Table 4.1 Genetic encoding of synaptic parameters for synapse encoding and node encoding. In the latter case the sign encoded on the first bit is applied to all outgoing synapses whereas the properties encoded on the remaining four bits are applied to all incoming synapses. A: genetically determined controllers; B: adaptive synapse controllers; C: noisy synapse controllers.

properties of all its incoming synapses (consequently, all incoming synapses to a given node have the *same* properties).

Synapse encoding allows a detailed definition of the controller, but for a fully connected network of $N$ neurons the genetic length is proportional to $N^2$. Instead node encoding requires a much shorter genetic length (proportional to $N$), but it allows only a rough definition of the controller. In recent work Floreano and Urzelai [67] showed that the evolutionary adaptive approach does not need a lengthy representation because the actual weights of the synapses are always shaped at run-time by the genetically specified rules. However, this is not possible in the traditional approaches where it is necessary to assign good initial weights to the controller. Therefore, the experiments reported here compare evolution of genetically-determined networks using synapse encoding with evolution of adaptive networks using node encoding.

What is encoded on the remaining four bits depends on the evolutionary condition chosen, namely:

1. *Genetically-determined*: 4 bits encode the synaptic strength. This value is constant during "life".

2. *Adaptive synapses*: 2 bits encode 4 adaptive rules and 2 bits the learning rate. Synaptic weights are always randomly initialized at the beginning of an individual's life and then updated according to their own adaptation rule.

3. *Noisy synapses*: 2 bits encode the weight strength and 2 bits a noise range. The synaptic strength is genetically determined

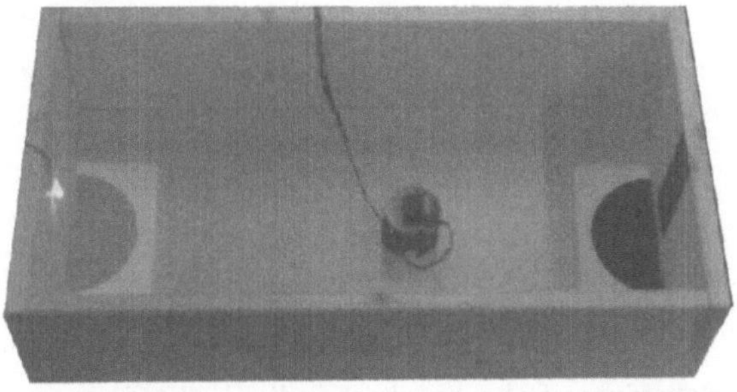

Figure 4.16 A mobile robot Khepera equipped with a vision module gains fitness by staying on the gray area only when the light is on. The light is normally off, but it can be switched on if the robot passes over the black area positioned on the other side of the arena. The robot can detect ambient light and the color of the wall, but not the color of the floor.

at birth, but a random value extracted from the noise range is freshly computed and added after each sensory motor cycle. This latter condition is used as a control condition to check whether the effects of Hebbian adaptation (condition above) are equivalent to random synaptic variability.

### 4.7.2 A Sequential Task

This set of experiments was designed to compare the performance of evolutionary adaptive controllers with respect to the evolution of synaptic weights and the evolution of noisy synapses in a sequential task that is complex enough to require non-trivial solutions. A mobile robot Khepera equipped with a vision module is positioned in the rectangular environment shown in Figure 4.16. A light bulb is attached on one side of the environment. This light is normally off, but it can be switched on when the robot passes over a black-painted area on the opposite side of the environment. A black stripe is painted on the wall over the light-switch area. Each individual of the population is tested on the same robot, one at a time, for 500 sensory motor cycles, each cycle lasting 100 ms. At the beginning of an individual's life, the robot is positioned at a random position and orientation and the light is off.

The fitness function is given by the number of sensory motor cycles

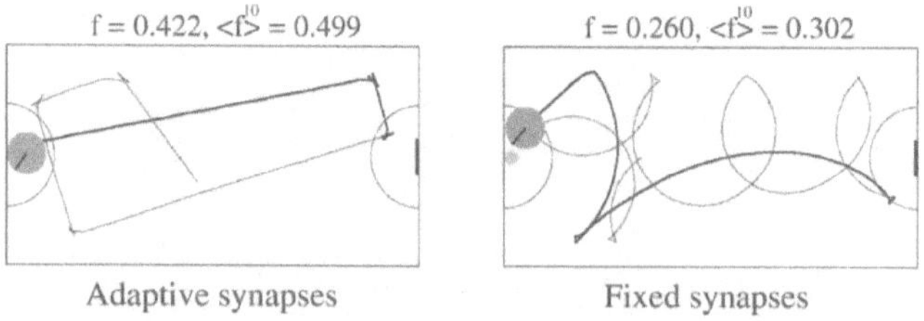


Figure 4.17 Behaviors of the two best individuals (from last generation) with adaptive synapses and node encoding (*left*) and with genetically-determined synapses and synapse encoding (*right*). When the light is turned on, the trajectory line becomes thick. The corresponding fitness value is printed on the top of each box along with the average fitness of the same individual tested ten times from different positions and orientations.

spent by the robot on the gray area beneath the light bulb *when the light is on* divided by the total number of cycles available (500). In order to maximize this fitness function, the robot should find the light-switch area, go there in order to switch the light on, and then move towards the light as soon as possible, and stand on the gray area. Since this sequence of actions takes time (several sensory motor cycles), the fitness of a robot will never be 1.0. Also, a robot that cannot manage to complete the entire sequence will be scored with 0.0 fitness.

A light sensor placed under the robot is used to detect the color of the floor—white, gray, or black— and passed to a host computer in order to switch on the light bulb and compute fitness values. The output of this sensor is *not* given as input to the neural controller. After 500 sensory motor cycles, the light is switched off and the robot is repositioned by applying random speeds to the wheels for 5 seconds. The experiments have been carried out in simulations sampling sensor activation and adding 5% uniform noise to these values [144]. All experimental conditions have also been repeated on the physical robot and yielded similar results [67].

The fitness results show that individuals with adaptive synapses and node encoding are much better than individuals with genetically-determined synapses and synapse encoding in that:

1. Both the fitness of the best individuals and of the population report higher values (0.6 against 0.5). The performance difference measured on best individuals of the last generation is statistically significant.

2. They reach the best value obtained by genetically-determined individuals in less than half generations (40 against more than 100).

Individuals with genetically-determined synapses and node encoding never managed to complete the task in any of the ten simulations performed.

Figure 4.17 shows the behaviors of the two best individuals evolved with adaptive synapses and node encoding (left) and with genetically-determined weights and synapse encoding (right). In both cases individuals aim at the area with the light switch and, once the light is turned on, they move towards the light and remain there. The better fitness of the adaptive controllers (given on the top of each box, see figure caption) is given by straight and faster trajectories showing a clear behavioral change between the first phase where they go towards the switching area and the second phase where they become attracted by the light. Instead, genetically-determined individuals always display the same looping trajectories around the environment with some attraction towards the stripe and the light. This minimalist behavior that depends on invariant geometrical relations of the environment gives them a chance to accomplish the task but with a lower performance.

#### From Simulations to Real Robots

One way of measuring the adaptive abilities of evolved controllers is to transfer them from simulated to real robots. Since physical robots and environments often have characteristics different from simulations, solutions evolved in simulation typically fail when tested on real robots. The best individuals of the last generation for each of the 10 populations evolved in simulation have been transferred on a physical Khepera robot. The performance of adaptive individuals is less affected by the transfer to the physical environment than genetically-determined individuals. Individuals with noisy synapses are not affected by the transfer because their behavior is always random and not effective in both simulated and physical environments.

Some performance loss in adaptive individuals is caused by the fact that in some cases the robot performs looping trajectories around

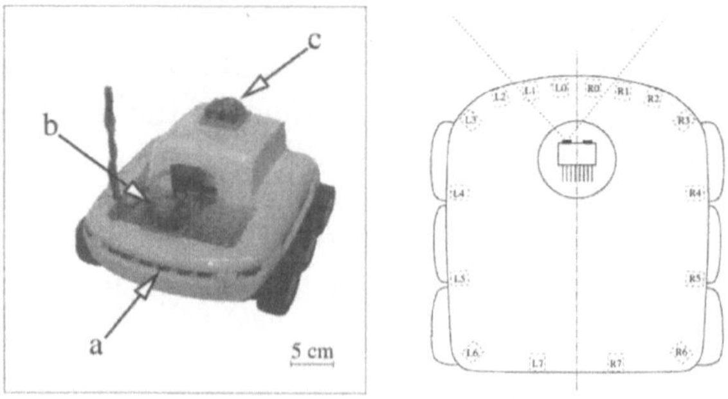


Figure 4.18 The Koala robot used in the experiments. Infrared sensors (a) measure object proximity and light intensity. The linear vision module (b) is the same as used in the experiments with the Khepera robot. The localization module (c) provides the position of the robot at every time step. Right figure shows the sensory layout of the Koala robot. Only 8 equally-spaced sensors are selected as input to the network.

the fitness area without coming to rest on it. Instead, the two major reasons for failure of genetically-determined individuals are that they often get stuck on the walls and they cannot manage to move efficiently towards the light. These failures are due to differences between simulated and real sensors. Since the weights are fixed, genetically-determined individuals cannot accommodate these changes as adaptive individuals do.

#### Crossplatform Evolution

Cross-platform transfer is a very useful feature but it is very difficult to transfer a control system across different robots without changes. One may develop (or evolve) control systems for a desktop sturdy robot like the Khepera and then download them to larger and consequently more fragile robots. Obviously, the two robots must share some characteristics, such as type of sensors and actuators used, that allow a suitable interfacing of the control system. In previous work it was shown that this can be achieved by using incremental evolution of genetically-determined networks [65]. However, even for a simple reactive navigation behavior it took an additional 20 generations to re-adapt to the new robot.

Here the adaptive properties of the evolutionary adaptive strategy are tested by transferring on a physical Koala robot (Figure 4.18, left) the best individuals of the last generation evolved in each of

Figure 4.19 A mobile robot Koala equipped with a vision module gains fitness by staying near the lamp (right side) only when the light is on. The light is normally off, but it can be switched on if the robot passes near the black stripe (left side) positioned on the other side of the arena. Position of the robot is controlled by an external positioning system and passed to the computer in order to control the light and to compute the fitness.

the 10 simulations of the experiment presented in Section 4.7.2. The Koala robot has six wheels driven by two motors (one on each side) and 16 infrared sensors (Figure 4.18, right) with a different and stronger detection range. A mobile robot Koala equipped with a vision module is positioned in the rectangular environment shown in Figure 4.19. As in the previous experiment with the Khepera robot, the Koala robot must find the light-switching area, go there in order to switch the light on, and then move towards the light as soon as possible and stay there in order to score fitness points.

An external positioning system emitting laser beams at predefined angles and frequencies is positioned on the top of the environment and the Koala robot is equipped with an additional turret capable of detecting laser and computing in real-time the robot displacement. This information is used in order to control the light and to compute the fitness. The performance of adaptive individuals is not affected by the transfer from the Khepera robot to the Koala robot, whereas genetically-determined individuals report a significant fitness loss. Individuals with noisy synapses are not affected by the transfer because their behavior is always random and not effective in both Khepera and Koala robots.

Individuals evolved in simulation for the Khepera robot display a satisfactory behavior when tested on the Koala robot. They correctly approach the light-switching area and they are clearly attracted by light (Figure 4.20, left). As in the case of a real Khepera robot,

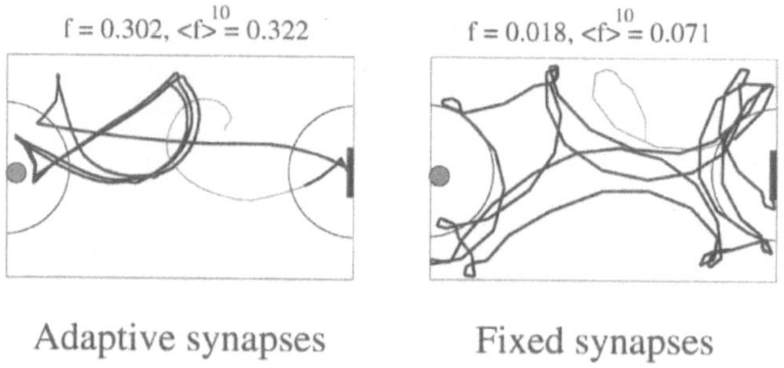


Figure 4.20 Behaviors of individuals with adaptive synapses (left) and genetically-determined synapses (right). Individuals belong to the last generation evolved in simulation for the Khepera robot.

once it has arrived under the light the Koala robot moves around the fitness area while remaining close to it until the testing time is over.

On the other hand, genetically-determined individuals (right) perform spiralling trajectories around the environment and do not display any attraction by the black stripe or the light. They eventually manage to pass through the light-switching area, turn the light on, and occasionally score fitness points passing through the fitness area. In several cases, genetically-determined individuals get stuck on the walls of the environment (behaviors not shown). Individuals with noisy synapses score a low performance because their strategy is based on random navigation.

### Conclusions

The experiments presented here show that evolution of adaptive synapses provides better adaptation capabilities than standard evolution of synaptic weights in the transfer from simulations to physical robots and across different robotic platforms. Evolutionary adaptive controllers can autonomously modify their synaptic weights and behavior online to the new environmental conditions without requiring additional evolution or ad-hoc manipulation of the evolutionary conditions.

The evolutionary technique proposed in [225] represents a significative step forward towards making evolutionary robotics applicable to real-world applications of autonomous robotics. In scenarios like those, for example, of robots probing an asteroid surface or robots interacting with a handicapped person it is impossible to evolve the control system on the spot (not even incrementally). However, one might reproduce the working conditions in the laboratory to some degree of approximation and evolve the adaptive controller in there.

The controller would then be transferred on the final robot and let free to adapt to actual working conditions in a few seconds. This adaptive strategy will also be useful for evolving more complex and powerful control architectures. In current methods there is a trade-off between the complexity the genotype/phenotype mapping and the evolvability of such systems which is partly due to the fact that the phenotype largely depends on genetic instructions. By evolving the adaptive characteristics along with other high-level parameters (position and type of nodes, e.g.) of the controller, one may obtain simpler genetic encodings and a higher tolerance to mutations. This would make the evolved controllers more viable, add neutrality to the genetic landscape, and ultimately improve evolvability.

Chapter 5

# Evolutionary Design of Fuzzy Systems

## 5.1 Introduction

ONE OF THE REASONS for the success of fuzzy logic is that the linguistic variables, values, and rules allow the engineer to seamlessly translate human knowledge into systems that work. What is a strength in some cases, however, is a weakness in others. If expert knowledge is not available, there is no ready made recipe to put together a fuzzy system from scratch, as is the case with more conventional techniques. This is where evolutionary algorithms come into play.

The idea of using evolutionary algorithms to design fuzzy systems is relatively recent. The first attempts in this direction were aimed at the synthesis and optimization of fuzzy controllers with genetic algorithms and were undertaken at the beginning of the 1990s by C. L. Karr [110] and P. Thrift [221].

Similar work was later carried out by Michael Lee and Hideyuki Takagi [123, 124] at the University of California, Berkeley and by Cezary Janikow [106] at the University of Missouri, St. Louis.

A common trait among these approaches is that they use very simple shapes for the membership functions (i.e. triangular or trapezoidal), so that these can be compactly encoded by two or three parameters, and they assume that the number of fuzzy domains for each variable is given and fixed. Furthermore, all possible rules, obtained as a Cartesian product of the input variables, are taken into account and their inclusion in the rule set of a controller is usually marked by a bit in the genetic encoding.

These simplifications greatly reduce the degrees of freedom in the search for an optimal controller while guaranteeing a sufficient amount of generality. The design process, however, is reduced to a parameter optimization problem in the framework of a rigid structure. In addition, the structure and implementation of the controllers is tailored for the convenience of the evolutionary algorithm.

Although the results obtained thus far are promising and this approach might prove to be entirely satisfactory from the practical point of view, it is not applicable to the case in which a *language* for the design of fuzzy controllers and a hardware architecture for fuzzy inference are given and a fuzzy controller has to be designed upon them. The availability of a dedicated fuzzy control processor, W.A.R.P., prompted Tettamanzi [219] to explore an alternative approach to the synthesis and optimization of fuzzy controllers using evolutionary algorithms. This approach could be described as *bottom-up*, in the sense that it starts from an existing fuzzy implementation from which it tailors a design technique rather than defining a suitable implementation given a design technique.

Besides control, another area of research is *data mining*, where evolutionary algorithms are used to optimize queries. This optimization task becomes particularly interesting when queries are vague, database indexing is fuzzy, and the data themselves are uncertain [200].

Further applications have been envisaged: for example using evolutionary techniques to tune a fuzzy image compression algorithm [25].

A broader notion of a fuzzy system includes not only program modules based on fuzzy IF-THEN rules, but also non-rule-based algorithms and systems that can make use of the fuzzy concepts of linguistic variables and values, membership functions, fuzzy logical operators, and fuzzy numbers and arithmetic.

Given the abstract task of designing a fuzzy system, there are several ways to carry it out by using evolutionary algorithms. As always, it is a matter of degree; the extent to which the designed fuzzy system depends on evolution can be mild or strong.

On the mild side, the overall structure of the fuzzy system may be evident from the beginning, and all that is required from an evolutionary algorithm is the tuning of some parameters. On the strong side, everything, from the structure to the parameters of the fuzzy system may be the object of the evolutionary search. A parallel can be traced with evolutionary design of neural networks (see Chapter 4), where an evolutionary algorithm may be used to find the weights in a fixed topology or to determine the network topology.

## 5.2 Evolutionary Design of Fuzzy Rule-Based Systems

FUZZY RULE-BASED systems are by large the most popular fuzzy systems studied and used in practical applications, and particularly in control. Therefore, it comes as no surprise that fuzzy logic controllers were the first fuzzy system to be designed by evolutionary algorithms.

Fuzzy rule-based systems include rules to direct the decision process and membership functions to convert linguistic terms into precise numeric values needed by the computer. The rule set can be gleaned from a human expert's knowledge and experience, and the membership functions are chosen by the system developer to represent the human expert's conception of the linguistic terms.

The selection of high-performance membership functions is commonly the most time-consuming phase of fuzzy system development. Changes in the membership functions alter the performance of the system because they determine the contribution each rule makes to the choice of the system's output. Thus, the performance of a fuzzy system is directly related to the developer's choice of membership functions.

For this reason genetic algorithms were in the first place applied to membership function determination and tuning. This reduces the development time required to design the fuzzy logic controllers substantially, although the intervention of an engineer is still required.

### 5.2.1 Membership Function Tuning

Procyk and Mamdani [180] and Shao [208] had proposed an iterative procedure for modifying membership functions, but, in general, little had been done to develop techniques for choosing membership functions that improve fuzzy system performance, before evolutionary algorithms began to be applied to that task.

The central issue when designing an evolutionary algorithm for membership function tuning is the encoding. This in turn depends on the assumptions that are made *a priori* about the shapes of membership functions.

### Isosceles Triangular Membership Functions

Karr [110, 109] applied a standard genetic algorithm to the problem of learning membership functions by imposing two important restrictions. First of all he constrained all membership functions to be isosceles triangles, with the exception of the two "extreme" membership functions, which are right triangles with their maximum corresponding to the limiting value of the variable definition interval. The second restriction is that the number of membership functions for each variable is predetermined and fixed.

The membership functions of a fuzzy controller are encoded as a finite-length bit-string. The parameters encoded are the anchor points that locate the triangles defining the linguistic variables. The "extreme" linguistic variables, like *negative big* and *positive big* require the definition of only one anchor point, since one side of the triangle is fixed at the limiting values of the variables. Two anchor points are required to describe the "interior" triangles, defining linguistic variables like *zero* or *negative small*. Only two anchor points are required because interior triangles are constrained to be isosceles.

### Tuning Membership Functions for the TSK Model

A similar approach is taken by Lee and Takagi [123], who adopt the Sugeno fuzzy model (cf. Chapter 3, Section 3.6.5) as the target fuzzy rule-based system.

Membership functions are triangular and parameterized by left base, right base, and distance from the previous center point (the first center is given as an absolute position). These three parameters are encoded as 8-bit numbers. The evolutionary algorithm is a standard genetic algorithm with bit-string representation.

A maximum number of membership functions per input variable is pre-determined, but the actual number of membership functions is implicitly determined by considering only those membership functions whose center points lie within the definition interval of the relevant variable.

### Other Types of Membership Functions

It is obvious that, as far as membership functions are defined by means of some parametric shape or function, the same basic approach can be adopted, and membership function tuning becomes in fact a parameter optimization problem, possibly with constraints.

Examples of parametric types of membership functions are: Gaussian, bell-shaped, trapezoidal, and sigmoidal.

### Fully Overlapped Membership Functions

Perhaps the most compact representation of triangular membership functions is obtained when adjacent membership functions are restricted to fully overlap: that is, the center of the previous membership function coincides with the left base point of the next [122]. The first membership function of each variable may also be constrained to have its center resting at the lower boundary of the variable definition interval.

By making these restrictions, only $m-1$ membership function centers need to be specified to represent $m$ membership functions. Furthermore, a differential encoding may be used for center points: instead of giving their absolute position, the representation may specify the distance from the center of the previous membership function.

## 5.2.2 Evolution of Rules

There are several encoding strategies suited for the evolutionary design of rules for fuzzy rule-based systems.

### Tabular Encoding

Thrift [221] considers a fuzzy rule base in tabular form. Let $x_1, \ldots, x_N$ be the input variables to the system. A fuzzy partition into $m_i$ domains is assumed to be defined on each variable $x_i$, $i = 1, \ldots, N$. The details of fuzzy partitions are described by membership functions for all the domains. An exhaustive set of rules for such a system can be represented as a multi-dimensional $m_1 \times m_2 \times \ldots \times m_N$ matrix $R$ , whose entries are linguistic values for the output variable $y$. This table representation is particularly convenient for simple systems having two input variables and one output variable. In that case a rule of the form

$$\text{IF } x_1 \text{ is } A_j \text{ AND } x_2 \text{ is } B_k \text{ THEN } y \text{ is } C \qquad (5.1)$$

translates into $C$ being the entry at row $j$, column $k$. Entries in the table can be blank, indicating the absence of a rule with the corresponding antecedent (IF-part).

Thrift then proposes to directly encode a rule base with a matrix, whose entries are fuzzy set labels over the output variable domain, plus a special *blank* value standing for *no output*. A linear genotype is formed from such a matrix by writing all the entries rowwise.

Standard mutation and crossover operators can act on these strings. However, Thrift uses a mutation operator that changes a label to the one corresponding to the previous or next sub-domain in the fuzzy partition or to the blank label. If the label is already blank, it chooses a non-blank label at random.

### Evolving Rules for the TSK Model

A substantially analogous approach is taken by Lee and Takagi [123], who adopt the TSK fuzzy model [216], whereby the consequent part of a fuzzy rule is a linear combination of the input values, as in Equation 3.56.

Lee and Takagi allot one gene in the genotype for each possible rule (with $N$ input variables and $m$ membership functions per variable, there are exactly $m^N$ possible rules). Such a gene encodes the $N+1$ weights $w_0, w_1, \ldots, w_N$, which determine the linear combination of the inputs for that rule, as 8-bit numbers.

The actual number of rules is implicitly represented by the combination of domain boundary knowledge and membership function center points. Rules containing, in their antecedent part, at least one membership function whose center point lies without the definition interval of the relevant variable are neutralized and can be eliminated from the rule base.

### Rule IDs

An alternative to the tabular encoding consists in assigning each possible rule a distinct integer number, the rule ID. A rule-base can then be encoded just by a list of rule IDs. This method is proposed in [122].

Assuming that $m$ membership functions are defined for each of $N$ input variables, as well as for a single output variable, the number of possible antecedent parts is $m^N$. A rule consists of one of the $m^N$ possible antecedent parts and any of the $m$ possible consequent parts. Therefore, a total of $m^{N+1}$ distinct rules can be formed. Possible rules can be enumerated and assigned IDs from 0 to $m^{N+1} - 1$. Such IDs can be encoded using $\lceil (N+1) \log_2 m \rceil$ binary digits.

Given a rule ID $d$, the corresponding rule can be found by means of the following procedure:

1. $i \leftarrow 1$;
2. the membership function for variable $x_i$ is $A_{ij}$, where $j = (d \bmod m) + 1$;
3. $d \leftarrow \lfloor d/m \rfloor$;
4. $i \leftarrow i+1$; if $i \leq N$ go back to 2; otherwise, go to the next step:
5. the output membership function is $B_d$.

This method can easily be extended to rule bases with more than one output variable.

### The Fuzzy Classifier System

The fuzzy classifier system, proposed in 1991 by Valenzuela-Rendón [228], merges the ideas behind classifier systems (see Chapter 1, Section 1.4) and fuzzy controllers. It has credit-assignment mechanisms which resemble those of common classifier systems, but with a fuzzy nature, and employs an evolutionary algorithm to evolve its rules.

The fuzzy classifier system operates over three types of variables: input, internal, and output. Input variables can be used in rule antecedents only, output variables in consequents only, whereas internal variables can be referenced to both in antecedents and in consequents.

Fuzzy rules are represented as binary strings encoding the membership functions of the fuzzy sets involved in the fuzzy rule. For each variable, $n$ sub-domains are pre-defined so that their membership functions span the entire definition interval, with their peaks equally spaced. The number $n$ is set by the user according to the precision desired. The number and shape of membership functions is fixed and not subject to evolution.

The binary string encoding for a rule

$$\text{IF } x_1 \text{ is } A_1 \text{ AND} \ldots \text{AND } x_N \text{ is } A_N \text{ THEN } y \text{ is } B \tag{5.2}$$

is of the form

$$\langle x_1 \rangle : \langle A_1 \rangle, \ldots, \langle x_N \rangle : \langle A_N \rangle \mapsto \langle y \rangle : \langle B \rangle \tag{5.3}$$

where $\langle X \rangle$ is the binary label of object $X$.

In a clause $\langle x \rangle : \langle A \rangle$, the predicate part $\langle A \rangle$ is binary coded so that the number of bits is the number of fuzzy sub-domains defined over variable $x$. A "1" indicates that the corresponding fuzzy sub-domain is part of the clause. More than one sub-domain can be part of the clause, this meaning that $x$ is to be in the union of the sub-domains selected.

For example, the rule

$$0 : 110 \mapsto 1 : 001 \tag{5.4}$$

defined for a system having one input variable $x_0$ and one output variable $x_1$, partitioned into the three linguistic values *low*, *medium*, and *high*, is to be interpreted as "IF $x_0$ is *low* or *medium* THEN $x_1$ is *high*".

The operation of a fuzzy classifier system is similar to that of a common classifier system, but classifiers are replaced by fuzzy rules. The basic cycle is as follows:

1. The input unit receives input values, encodes them into fuzzy messages, and adds these to the message list.

2. The rule base is scanned to find all rules whose conditions are satisfied by the messages in the message list.

3. The message list is erased.

4. All matched rules are fired, and the messages produced are stored in the message list.

5. The output unit detects the output messages, and removes them from the message list.

6. In the output unit, output messages are decomposed into minimal messages.

7. Minimal messages are defuzzified and transformed into output values.

8. Payoff from the environment and rules is transmitted through the messages to the rules.

Figure 5.1 is a block diagram of a fuzzy classifier system.

The values taken by variables are broadcast to the classifier as *messages*. Each rule will verify whether its conditions are satisfied

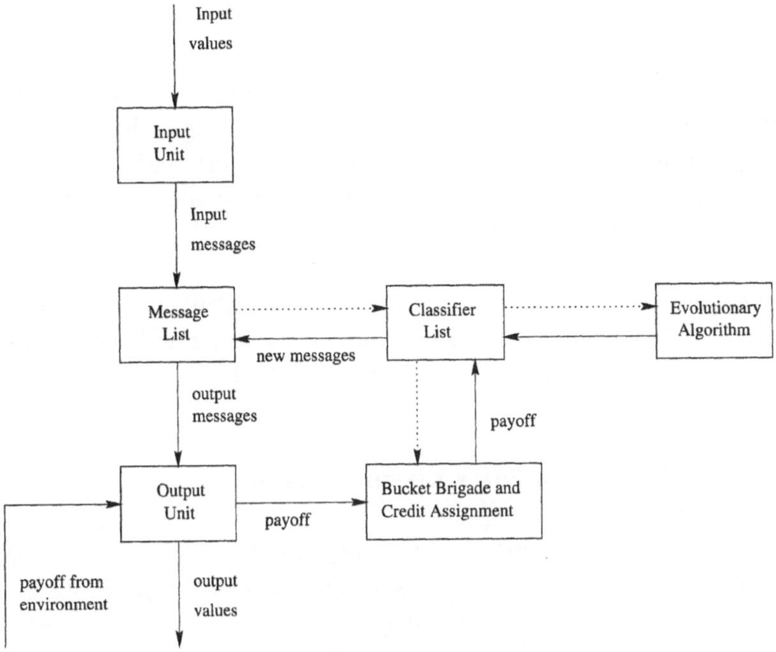


Figure 5.1 A block diagram of the fuzzy classifier system.

by the messages and, if so, post a new message according to its consequent.

The process of a rule verifying whether its conditions are satisfied by a message can be viewed in two ways:

- the rule receives crisp values and matches them against the relevant fuzzy sets in its antecedent;
- the rule receives fuzzy values and performs an exact match with its conditions.

For the sake of performance, Valenzuela-Rendón's system implements the second approach: the input units fuzzifies inputs into fuzzy messages. Each message has an associated activation given by the degree to which the crisp value belongs to the fuzzy set represented by the message. Fuzzy rules match these messages and produce new messages with activation corresponding to the degree to which their conditions are satisfied. The fuzzification of inputs is accomplished by creating *minimal messages*, one for each fuzzy set defined over a variable. A minimal message is a message $\langle x \rangle : \langle A \rangle$ having a single "1" in the $\langle A \rangle$ part.

According to the calculus of fuzzy rules, the satisfaction of a condition equals the maximum activation of the messages that match it (fuzzy disjunction). The global activation of a rule, and therefore of the message it produces, equals the minimum of the satisfaction of its conditions (fuzzy conjunction).

The defuzzification is accomplished by the output unit by decomposing each output message into minimal messages and then applying the center of gravity defuzzification method.

### Credit Assignment in Fuzzy Classifier Systems

The output value, produced by the output unit, is evaluated against the problem the fuzzy classifier system has to solve or the task it must carry out, and a *payoff* is determined accordingly. The output unit distributes this payoff through the minimal messages to the rules that produced the output messages. The payoff is distributed in such a way that rules that contributed more to the output taking the actual value receive a larger share of the payoff. In this way, rules that are directly involved in producing correct outputs increase their fitness.

Payoff to other rules that do not directly produce outputs, but post messages that allow other rules to fire and receive payoff, is distributed following the "bucket brigade" algorithm of standard classifier systems (cf. Chapter 1, Section 1.4).

According to the "bucket brigade" algorithm, matched rules *bid* a small portion of their fitness for the right to fire. Firing rules pay their bids to those that posted the messages that allow them to fire. In this way, a competitive economic system is established in which rules that produce or help produce good outputs have their fitness increased, and others have their fitness decreased. A bias against rules that do not participate in the competition is achieved through a sort of taxation, whereby all rules have a small portion of their fitness deducted at every generation.

### 5.2.3 Evolution of Membership Functions and Rules

The simplifications introduced by the various techniques illustrated in the previous sections greatly reduce the degrees of freedom in the search for an optimal fuzzy rule-base, while guaranteeing a sufficient amount of generality. The design process, however, is reduced to a parameter optimization problem in the framework of a rigid structure. In addition, the structure and implementation of the fuzzy systems is tailored for the convenience of the evolutionary algorithm.

Although the results obtained thus far are promising and those approaches can be entirely satisfactory from the practical point of view, it is not applicable to the case in which a *language* for the design of fuzzy systems and a predetermined architecture for fuzzy inference are given and a fuzzy rule-base has to be designed upon them.

An alternative approach could be described as *bottom-up*, in the sense that it starts from an existing fuzzy implementation from which it tailors a design technique rather than defining a suitable implementation given a design technique.

An example of this type of approach is given in [177], where an evolutionary algorithm is designed for the automatic synthesis and optimization of fuzzy controllers for the Weight Associative Rule Processor (W.A.R.P.) manufactured by SGS-Thomson Microelectronics.

The W.A.R.P. is a chip for fuzzy control applications manufactured by SGS-Thomson Microelectronics. The architecture of the W.A.R.P. chip arose from the need for an integrated structure with high inferencing performance and flexibility. W.A.R.P. supports antecedent membership functions with any shape, up to 256 rules, each comprising up to four antecedent clauses and one consequent clause; up to 16 input variables can be handled, described by up to 16 distinct membership functions each, while as many as 128 membership functions can be defined for output variables.

Writing an evolutionary algorithm that is able to exploit the versatility of such a tool in order to synthesize and optimize fuzzy rule sets for control problems posed an interesting methodological challenge. An encoding that directly maps to the machine level of W.A.R.P.'s programming language was defined, and specialized mutation and recombination operators were designed. Every parameter involved in the design, from the number of rules and membership functions to their structure and shape, is discovered by the evolutionary process; the only constraints for the evolved controllers are those imposed by the processor's hardware implementation.

### Encoding

W.A.R.P. programs are encoded in three main blocks: a set of membership functions for the input (or status) variables, a set of symmetric membership functions, represented by means of area-center of mass pairs for the output (or control) variables, and a set of rules.

A single input variable membership function is defined by a list of 32 points; a point is a pair of fixed point numbers fitting into a byte. Output variable membership functions are assumed to be symmetrical, and thus can be described using just two parameters: their area and center of mass. A rule is a list of up to four conjoint antecedent clauses (the IF part) and a consequent clause (the THEN part). A clause is represented by a pair of indexes referring respectively to a variable and to one of its fuzzy subdomains, i.e., a membership function.

W.A.R.P.'s hardware implementation requires that, for each output variable, two membership functions exist and be used in at least one rule. This requirement is enforced in the algorithm by the convention that the first rules have fixed consequent sides; moreover, the first two membership functions of every output variable coincide respectively with the minimum and maximum of the relevant definition interval, with null area (that is, they represent crisp extreme values for that variable).

### Initialization

The population can be seeded either with hand-written or otherwise already existing W.A.R.P. programs or with new random ones. A new individual is created according to the following algorithm:

1. Each input variable must have at least one domain; the number of additional domains is determined by sampling from a truncated exponential distribution with mean 3;

2. The shapes of the membership functions for the input variables are determined by random extraction of a centroid $C$; then, two sigmoids with their maximum in $C$ are drawn on both sides to yield an asymmetric bell-shaped curve, whose values in the extremes of the interval $[a, b]$ of the relevant input variable are null:

$$\mu(x) = \begin{cases} \sin \frac{\pi(x-a)}{2(C-a)}, & x < C; \\ \sin \frac{\pi(x-b)}{2(b-C)}, & x \geq C; \end{cases} \tag{5.5}$$

3. At least two output membership functions have to be present for each output variable; the number of additional domains for each output variable is determined by sampling from a truncated exponential distribution with mean 3.

4. The centers of mass for the output variables are randomly extracted in the range of the relevant variable; the areas are extracted at random such that they correspond to a triangular membership function whose base is entirely contained in the range of the variable;

5. At least two rules have to be present, using at least two different output domains; the number of additional rules in the rule base is determined by sampling from a truncated exponential distribution with mean 6;

6. the rules are generated according to the following algorithm:

   (a) for each input variable, a fair coin is flipped to decide whether to include it in the antecedent part, not exceeding four variables;

   (b) for each selected input variable, a domain is extracted among those defined;

   (c) an output variable and its domain are extracted for the consequent part of the rule.

**Recombination**

The recombination operator is designed to preserve the syntactic legality of W.A.R.P. programs. A new W.A.R.P. program is obtained by combining the pieces of two parent programs. Each rule of the offspring program can be inherited from one of the parent programs with probability 1/2. When inherited, a rule takes with it to the offspring program all the referred domains with their membership functions. Other domains can be inherited from the parents, even if they are not used in the rule set of the child program, to increase the size of the offspring so that their size is roughly the average size of the parents.

### Mutation

Like recombination, mutation also produces only legal W.A.R.P. programs. Mutation can result in one or more of the following changes, with probability given by the mutation rate, $p_m$, identical and independent for each component of the genotype:

- a new domain with a random membership function is added to an input variable;
- a domain whose membership function is not used in the rules is removed from an input variable;
- a membership function is perturbed as follows:
    1. a point $X$ in the range of the relevant variable is extracted at random with uniform probability;
    2. a new value $Y \in [0, 1]$ for the membership function in $X$ is extracted, again with uniform probability;
    3. the spread $\sigma$ of the perturbation is extracted from an exponential distribution with mean one quarter of the variable interval;
    4. the old membership function $\mu$ is modified into a new one $\bar{\mu}$ such that $\bar{\mu}(X) = Y$,

    $$\bar{\mu}(x) = \lambda Y + (1 - \lambda)\mu(x), \qquad (5.6)$$

    where

    $$\lambda = \frac{1}{\left(\frac{X-x}{\sigma}\right)^2 + 1}. \qquad (5.7)$$

- a new domain, with random area and center of mass, is added to an output variable;
- a domain whose area and center of mass are not used in the rules is removed from an output variable;
- an area/center of mass pair is perturbed as follows:
    1. a standard deviation $\sigma$ for the perturbation is extracted from an exponential distribution;
    2. a new center of mass is extracted from a truncated normal distribution with mean the old center of mass and standard deviation $\sigma$;

3. a new area is extracted from a truncated exponential distribution with mean the old area, such that it corresponds to a triangular membership function entirely contained in the range of the relevant output variable;

- a new random rule is added to the rule set; the new rule is generated as follows:
  1. for each input variable, a fair coin is flipped to decide whether to include it in the antecedent part, not exceeding four variables;
  2. for each selected input variable, a domain is extracted among those defined;
  3. an output variable and its domain are extracted for the consequent part of the rule;
- a rule is removed from the rule set;
- a rule gets a random antecedent clause predicating an input variable not yet used added to it;
- an antecedent clause is removed from a rule;
- the predicate of an antecedent clause is modified by randomly extracting one of the domains defined for the relevant input variable;
- the predicate of the consequent clause of a rule is modified by randomly extracting one of the domains defined for the relevant output variable.

### 5.2.4 Evolving Fuzzy Controllers

The most widespread application of fuzzy rule-based systems lies in the area of control. Apart from the issues discussed above, which are general for all fuzzy rule-based systems, the definition of an appropriate fitness function is a critical (and not at all obvious) step in the design of an evolutionary algorithm for fuzzy controller design.

An acceptable controller is required to drive the controlled physical system into a target state (or subspace of states) in the shortest possible time, and then to keep it in that state for the longest possible time (ideally forever), while minimizing the energy needed to perform control actions, a goal that often coincides with the requirement of avoiding, to the maximum possible extent, oscillations.

It is clear that designing a fitness function to take all these objectives, which are partially conflicting, into account is not a straightforward task.

In order to assess the quality of a controller, a simulator of the physical system must be available. Evaluating evolved controllers directly on the real system is not feasible for a number of reasons:

- *time* — in many cases, it takes much less time to run a controller on a simulation than on the real system;
- *cost* — in most practical settings, devoting the real system to testing controllers for an evolutionary algorithm for a significant extent of time is just too expensive: for instance, stopping a production line in a factory for this kind of testing would translate into lack of production, energy consumption, material consumption, etc.;
- *security* — especially in the first stages of the evolutionary process, bad controllers might cause serious damage to the real system and to its surrounding environment: imagine for example the real system being a chemical reactor or a mobile robot.

Assuming that a simulator of the real system is available, another issue of capital importance is selecting a set of initial conditions on which controllers are to be tested. These initial conditions are called *test cases*. There are two fundamental approaches to this issue:

- determine a set of test cases that is representative of most situations the controller might have to confront in practice and then use the same cases to evaluate all controllers produced by the evolutionary algorithm;
- generate initial conditions at random according to a probability distribution that reflects the relative frequency of the various situations occurring in reality each time a controller has to be evaluated

Both approaches have advantages and drawbacks. In the first approach, there is no guarantee that an evolved controller that works well on the test cases will work equally well under other conditions, but if the set of test cases is thoughtfully chosen it might be reasonable to assume a certain degree of robustness. In the second approach, the fitness of an individual will be *noisy* and a series of special considerations will have to be applied in designing the evolutionary algorithm, but the robustness of the evolved controller will be higher.

As an example of the first approach, Karr [109] suggests having the evolutionary algorithm minimize a squared weighted distance from the target state, summed over some finite time. Mathematically, this objective function is described by the expression

$$z = \sum_{c=1}^{C} \sum_{t=0}^{T} \sum_{i=1}^{N} w_i (x_i(c,t) - x_i^*)^2, \tag{5.8}$$

where $N$ is the number of variables describing the state of the system to be controlled, $T$ is a limit time for simulation, and $C$ is the number of test cases used to try the controller. Time is treated as a discrete variable because in general simulation on a computer is performed in discrete steps, but the second summation in Equation 5.8 might be replaced by an integral when appropriate. The values $w_1, \ldots, w_N$ are weight factors compensating for the different scale and importance of state variables.

### Adaptive Fuzzy Control

Other considerations are in order where evolutionary algorithms are applied to the control of complex time-variant systems. These require adaptive fuzzy logic controllers, able to account for changes in the characteristic parameters before the changes cause the physical system to fail. In this case, an evolutionary algorithm has to be executed on-line, i.e., while the physical system and its controller are running, and in real-time.

The real-time requirement means that evolution must be quick enough to adapt to changes in the physical system before these changes lead to malfunction.

Small-population genetic algorithms have been theoretically shown to be more effective [78] and have proven to be efficient in non-stationary optimization problems [120].

The basic form of the objective function used to calculate the fitness of controllers in Equation 5.8 can be used in this case too, with a difference: in many applications it is unnecessary to consider a set of test cases; only the conditions at the current time must be considered.

As a matter of fact, the duration of the simulations takes up a greater importance in the case of on-line adaptation. If the simulations utilized to calculate the fitness of the controllers require too much time, the evolutionary algorithm will no longer be able to track changes before it is too late. On the other hand, the accuracy of the membership functions tuned by the evolutionary algorithm improves with longer simulations. Therefore, a balance must be struck between these two conflicting criteria.

## 5.3 Evolving Fuzzy Decision Trees

DECISION TREES provide one of the most popular methodologies for symbolic knowledge acquisition and representation. The acquired knowledge is structured in the form of a symbolic tree and a straightforward inference mechanism permits its use for classification purposes. Decision trees were made popular with Quinlan's ID3 [183], which was designed to work with symbolic domains for both features and decisions. If some of the features or decisions are defined on a continuous space, the decision tree approach creates potential problems.

That is why Janikow [105] proposed in 1993 to extend decision trees with fuzzy logic, obtaining fuzzy decision trees. "Crisp" decision trees are made of two major components: a procedure to build the symbolic tree, and an inference procedure for decision making. In fuzzy decision trees, knowledge is represented at three different levels: the symbolic tree structure, fuzzy sets, and inference procedures. He then showed how to use evolutionary algorithms to extract the knowledge from data by simultaneously operating at the first two of these levels [107].

### 5.3.1 The Partitioning Algorithm

Quinlan's ID3 partitioning algorithm serves as the basis for the fuzzy decision tree partitioning algorithm. The root of the decision tree contains all training examples. It is recursively split with the examples partitioned. The splitting stops when the examples contained by a node represent a unique decision, all attributes are used on a path or when some other user-specified criteria are met. The nodes at which the splitting stops become leaves and assume the decision of their examples. Node splitting is performed by selecting one of the attributes not yet used on its path: domain values of that attribute become the conditions leading to child nodes and the examples associated to the node are partitioned among its children, according to the condition they match. One of the most popular strategies for attribute selection is to maximize the information gain:

1. compute the information content at node $n$, given by

$$I_n = \sum_{d=1}^{D} \frac{N_d(n)}{N(n)} \log \frac{N(n)}{N_d(n)}, \tag{5.9}$$

   where $D$ is the number of decisions, $N(n)$ is the number of all training examples in node $n$, and $N_d(n)$ is the number of training examples in node $n$ having decision $d$;

2. for each attribute $a$ not appearing on the path to $n$, and for each of its domain values $a_i$, compute the information content $I_{n|a_i}$ in the node restricted by the condition that $a$ is $a_i$;

3. compute the information content

$$I_{n|a} = \sum_{i} q_i I_{n|a_i} \tag{5.10}$$

   based on selecting value $a_i$, where $q_i$ is the portion of examples at the child node matching the condition "$a$ is $a_i$" relative to all examples in $n$;

4. select the $a$ that maximizes the gain $I_n - I_{n|a}$.

Fuzzy decision trees consider attributes and decisions with numerical values and treat them as linguistic variables. The tree-building algorithm is an extension of Quinlan's ID3 partitioning algorithm. The main difference is that the examples are counted using fuzzy measures. Therefore,

$$N_d(n) = \sum_{e \in E} w_n(e) \mu_d(e), \tag{5.11}$$

where $E$ is the set of all examples, $w_n(e)$ is the degree of membership of example $e$ in node $n$, and $\mu_d(e)$ is the degree to which $e$ has decision $d$.

Given this tree-building algorithm, optimization of fuzzy decision trees can proceed by adjusting the definitions of linguistic variables (i.e., the membership functions defined on the numerical attributes) to minimize the information content of the data. This optimization, to be performed by an evolutionary algorithm, can be *static*, that is carried out prior to tree construction, and *dynamic*, that is performed while constructing the tree. The overall procedure is the following:

1. statically optimize the initial linguistic values (fuzzy sets) for all linguistic variables (attributes and decisions). Put the root node, containing all the training examples, in *OPEN*. Measure the time $T$ needed for the optimization (based on some criteria, e.g., CPU time, evaluation steps, generations). Set $\hat{A} = \emptyset$.

2. If $OPEN \neq \emptyset$, select the node $n$ having the highest number of training examples $N(n)$:

   (a) if $\hat{A} \neq \emptyset$, optimize the linguistic values of attributes not belonging to $\hat{A}$, allowing $T/\|\hat{A}\|$ optimization time;

   (b) compute $N_d(n)$, $N(n)$ and $I_n$;

   (c) remove $n$ from *OPEN*;

   (d) if the stopping criteria are not met, go to 2;

3. for each attribute $a$ not appearing in the path to $n$, compute $I_{n|a}$;

4. select for split the $a$ maximizing $I_n - I_{n|a}$:

   (a) if $a \notin \hat{A}$, optimize its linguistic values, allowing $T/\|\hat{A}\|$ optimization time; put $a$ into $\hat{A}$;

   (b) put each of $n$'s children into *OPEN*.

5. Go to 2.

### 5.3.2 The Optimization Algorithm

The evolutionary algorithm used to perform the static and dynamic optimization in Steps 1 and 2.a of the tree-building algorithm has to process many constraints, resulting both from various requirements imposed on the fuzzy sets and from the properties of such sets.

Instead of devising a representation that implements some of the constraints, Janikow explicitly utilizes all the constraints to reduce the search space. Trapezoidal membership functions, defining the linguistic values, are directly encoded by the $x$-coordinates of their four corners, and all equality and inequality constraints are placed into the evolutionary algorithm. This algorithm, GENOCOP, described in detail in [143], processes equality constraints to explicitly reduce the number of optimization variables by removing dependent variables. Then, inequalities are used by closed genetic operators, which produce only feasible offspring from feasible parents.

Let $x_1^{a,i}, x_2^{a,i}, x_3^{a,i}, x_4^{a,i}$ be the $x$-coordinates of the four corners of the trapezoid defining the $i$th linguistic value of attribute $a$. There are two inequality constraints:

- Trapezoidal constraints: $x_1^{a,i} \leq x_2^{a,i} \leq x_3^{a,i} \leq x_4^{a,i}$.
- Non-containment: $x_4^{a,i+1} > x_4^{a,i}$. This ensures that no fuzzy set range contains the range of another fuzzy set.

The most general equality constraints are the following:

- Overlap: $x_3^{a,i} + x_4^{a,i} = x_1^{a,i+1} + x_2^{a,i+1}$. This constraint applies to neighboring fuzzy sets and requires that there is an intermediate element between them which has a membership degree of $\frac{1}{2}$ to both sets or that, in other words, their membership functions intersect at a membership degree of $\frac{1}{2}$.
- Completeness: the leftmost and rightmost fuzzy sets must completely cover the left and right portions of the attribute range.
- Set symmetry: $x_2^{a,i} - x_1^{a,i} = x_4^{a,i} - x_3^{a,i}$. This ensures the trapezoids are symmetrical (it may not always be desired).
- Domain symmetry: $x_m^{a,i} = -x_m^{a,n_a-i}$, for $m = 1, \ldots, 4$, where $n_a$ is the number of domains for attribute $a$. This constraint is often used when the attribute range is centered around zero and the linguistic values on both sides of it are assumed to be symmetrical.

Of course, other more specific constraints may be formulated, depending on the particular nature of a given problem.

## 5.4 Evolving Fuzzy Queries

INFORMATION retrieval systems, which include but are not limited to databases, are widespread in every field and their use is of great and increasing economical relevance.

When queries are submitted to complex databases by users unfamiliar with the underlying representation model or with the retrieval operations, a special treatment of queries is often indicated, in particular to optimize them and improve the search process.

### 5.4.1 Information Retrieval Background

There are several indices allowing the measurement of the performance of information retrieval:

- *relevance* — the relevance to a query is the degree to which results provide an answer to that query;
- *precision* — the average relevance of the retrieved documents (the higher their relevance, the higher the precision): precision measures the degree to which only relevant documents are retrieved.
- *recall* — measures what fraction of the relevant documents is actually retrieved or, in other words, the completeness of the results.

Fuzziness comes in several ways in fuzzy information retrieval [200]:

- fuzzy indexing, whereby records in a database can be viewed as represented by vectors of weights corresponding to the index terms that describe their topic;
- retrieval status values computed through fuzzy sets operators;
- fuzzy queries and use of ranked output documents rather than retrieval of a single set.

Viewed as an extension of Boolean query models, fuzzy models take into account the two lacking features of partial relevance and weighting of queries. Outcomes of fuzzy information retrieval systems consist of documents in a search space, that are ranked according to their evaluated relevance. Therefore, in response to a query, a choice of most significant documents can be provided to the user, or a threshold on the retrieved status values may be used as a cut-off on the number of documents. It has been noticed that generation of ranked results increases user satisfaction and retrieval precision.

The first suggestion regarding the use of evolutionary algorithms in information retrieval dates back to 1988 and is due to M. Gordon [79]. Gordon introduced a genetic model of information retrieval for the improvement of document descriptions (for example sets of possibly weighted terms or keywords) incorporating user evaluation. Genotypes are bit strings that encode a description of a document or a query. Each bit corresponds to a single term or keyword. From the outcome of previous queries, the model determines how a document should have been described for better retrieval, and this information is used to improve the performance of the queries to come. A document has several complete descriptions. Each of them is evaluated against a given query according to the Jaccard score. The Jaccard score measures the similarity between two sets $A$ and $B$ as

$$s_J(A,B) = \frac{\|A \cap B\|}{\|A \cup B\|}. \tag{5.12}$$

Here, the two sets being compared are the set of terms in a document description and the set of terms appearing in the query. The fitness for the genetic algorithm is given by the average of the Jaccard score of a query with the desired documents.

### 5.4.2 Tuning Fuzzy Sets of Terms

Complementary to Gordon's approach to optimizing document descriptions, is Sanchez and Pierre's proposal for fuzzy query optimization [200]. Their approach consists in having an evolutionary algorithm search for the fuzzy set of terms (i.e., keywords) in the query that causes the greatest number of relevant documents to be retrieved. The genotype consists of an array of real numbers $0 \leq \mu_i \leq 1$, where $\mu_i$ is interpreted as the degree of membership of the $i$th term in the fuzzy query.

When searching large document spaces, it is often the case that the user is not completely satisfied with the documents that are retrieved at the first attempt. What happens then is that the user modifies or tunes the query several times according to the documents actually retrieved, until a satisfactory set of relevant documents is reached.

Therefore, the idea is to use relevance feedback from the user to assign a fitness to queries in the evolutionary algorithm, much in the spirit of the interactive genetic algorithm described in [34]. The user is asked to provide a list of terms and to formulate a compound query. The user has the possibility of assigning weights to individual terms.

The initial population is generated by assigning the degrees of membership $\mu_i$ based on the weights provided by the user or at random, if the user has not specified weights, for each individual.

Standard recombination, mutation, and selection operators are then applied to the population to evolve better fuzzy sets of terms for the given query.

Positive feedback is concerned with information from retrieved documents considered as relevant, negative feedback comes from documents considered not relevant, and mixed feedback makes use of relevant as well as non-relevant documents to tune queries. Positive feedback only is used for fitness evaluation, as it has been observed that it is more suited than negative or mixed feedback to adjusting queries.

### 5.4.3 Optimizing the Whole Query

A more sophisticated approach along the same lines is the one by Kraft and colleagues [119]. They treat fuzzy queries as a special case of a program, written in a language having the usual logic operators as functions and terms (keywords) and their degrees of membership (or importance) as the set of terminal symbols. Then they apply genetic programming (cf. Chapter 1, Section 1.5) to find the optimal form for the query.

Consider a series of terms, $t_1, t_2, \ldots, t_n$. An example of an unweighted Boolean query would be an expression like

$$(t_1 \wedge (t_3 \vee t_5)) \wedge (t_4 \vee t_2) \tag{5.13}$$

Expression 5.13 might be put in prefix LISP-like form,

$$\texttt{(AND (AND } t_1 \texttt{ (OR } t_3\ t_5\texttt{)) (OR } t_4\ t_2\texttt{))} \tag{5.14}$$

thus becoming a chromosome ready for genetic programming. Furthermore, fuzzy weights, or degrees of truth or importance, can be attached to each term, thus giving a fuzzy query of the form

```
(AND (AND (μ1 t1) (OR (μ3 t3) (μ5 t5)))
     (OR (μ4 t4) (μ2 t2))).
```

As the recombination operator, Kraft and colleagues use the standard crossover of genetic programming. For the mutation, they define a problem-specific operator. Mutation of the symbolic strings is straightforward: binary AND and OR can be changed, a unary NOT operator can be inserted, and a term can be added or deleted. Finally, the weight associated with a term can be modified. This mutation of the nonsymbolic components is performed by perturbing a weight $\mu$ of a small positive or negative value $\epsilon$ chosen at random. This gives a mutated weight $\mu' = \mu + \epsilon$.

As for the fitness, the most common measures of retrieval effectiveness are precision and recall. Thus, it is reasonable to express the fitness of a query as a function of its recall and precision. These two measures often tend to be inversely proportional, or conflicting, leading to difficulties. Kraft and colleagues, according to experimental evidence, suggest that a good definition for the fitness function is one mainly based on recall, with a factor added for precision:

$$f = \alpha \frac{\sum_d r_d \chi_d}{\sum_d r_d} + \beta \frac{\sum_d r_d \chi_d}{\sum_d \chi_d}, \qquad (5.15)$$

where $0 \leq r_d \leq 1$ is the relevance of document $d$ and $\chi_d \in \{0, 1\}$ is its retrieval (i.e., whether it is retrieved by the query or not).

Another possible definition of fitness is the one inspired by Salton's work on relevance feedback [196]. Let $n_{dc}$ be the number of conjuncts in a query and $c_j$ the $j$th conjunct in the query. Then, the fitness of the query is defined as

$$f = \alpha \sum_d r_d \sum_{j=1}^{n} \frac{s(d, c_j)}{n_{dc}} - \beta \sum_d (1 - r_d) \sum_{j=1}^{n} \frac{s(d, c_j)}{n_{dc}}, \qquad (5.16)$$

where $s(d, c_j)$ is a similarity measure of document $d$ to conjunct $c_j$. Note that $\alpha$ and $\beta$ are arbitrary weights that can be used to adjust the trade-off between the two terms.

The definition of Equation 5.15 has the advantage of being independent of the structure and form of the queries generated by the evolutionary algorithm. The alternative in Equation 5.16, on the other hand, has the disadvantage of requiring that the query be divisible into conjuncts, but has the advantage of being able to distinguish between queries that have identical retrieval behaviors.

## 5.5 Evolving Fuzzy Filters

AN EXAMPLE of the use of evolutionary algorithms to optimize a non-rule-based fuzzy system is presented in [25, 24] and [23].

An approach to image compression that makes use of fuzzy logic is the following: an image is divided into squares, and a fuzzy compression filter replaces each square in the original image with a single pixel in the compressed image; a fuzzy decompression filter then reconstructs the individual pixels of the decompressed image on the basis of the pixels in the compressed image.

The content of a square must be mapped into a single pixel $\mathbf{x}$ in the compressed picture. The transformation is of the form

$$\begin{aligned} x_C &= w\frac{\sum_i \sum_j \mu_{\mathbf{x}}(i,j) I_{ij}^C}{\sum_i \sum_j \mu_{\mathbf{x}}(i,j)} + \\ &+ (1-w)\sup_{i,j}\{\mu_{\mathbf{x}}(i,j) I_{ij}^C\}, \end{aligned} \tag{5.17}$$

for each component $C$, where $I$ is the source image and $\mu(\cdot)$ is a membership function defining the square.

Given a compressed image $X$, obtained as a result of the fuzzy compression algorithm, that is a collection of color values for each square in the source image, the color of a pixel in the decompressed image, $J_{ij}$, is given by the following expression, for each of the three color components $C$:

$$J_{ij}^C = \sum_{\mathbf{x}\in X} \pi_{\mathbf{x}}(i,j) x_C, \tag{5.18}$$

where $\pi_{\mathbf{x}}$ (which in general will be distinct from $\mu_{\mathbf{x}}$) is a membership function associated with square $\mathbf{x}$.

The success of this approach critically depends on the proper tuning of many parameters, defining the shapes of the membership functions utilized in the compression and decompression filters, which is accomplished by an evolutionary algorithm.

The technique mentioned above can be applied to signals of arbitrary dimensions. One-dimensional and two-dimensional signals, which are ubiquitous in a variety of practical applications, are of particular interest.

### 5.5.1 The Evolutionary Algorithm

#### Encoding

A membership function pair is encoded as an array of 22 reals in $[0,1]$, of which 11 parameterize the compression membership function and 11 parameterize the decompression membership function. The assumption is made that compression and decompression membership functions are symmetrical with respect to the center of their interval.

#### Mutation and Crossover

Mutation perturbs, with probability 0.1, a membership function as follows: the old membership function $\mu$ is modified into a new one $\bar{\mu}$ such that $\bar{\mu}(x) = \lambda Y + (1-\lambda)\mu(x)$, where $\lambda = (((X-x)/\sigma)^2+1)^{-1}$, $X$ and $Y$ have uniform distribution and $\sigma$ has exponential distribution with mean 0.5.

Uniform crossover with rate 0.6 operates on floating-point numbers. This means that each parameter in the offspring has equal chance of being inherited from either parent.

#### Selection

A membership function pair to be evaluated is used to compress and decompress a set of signals. Its fitness is a function of the average mean square error between the original and the final signals. An elitist exponential ranking selection is adopted, which assigns, to the $k$th individual in the ranking, a $1/2^k$ probability of being selected.

### 5.5.2 Results

Experiments have been performed with a population size of 100 individuals for 2000 generations. A typical pair of membership functions found by the algorithm is shown in Figure 5.2. Figure 5.3 shows the result of applying the algorithm with the membership functions of Figure 5.2 to a sample signal.

Figure 5.4 shows a typical result of applying the compression and decompression filters to a two-dimensional signal, for example a picture. Figure 5.4 shows the original picture and the relevant final result after passing it through a fuzzy compression filter with a ratio of 64 to 1.

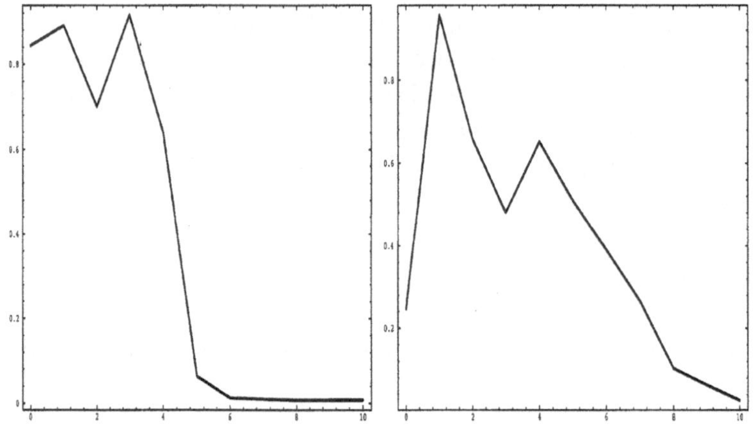

Figure 5.2 An example of compression/decompression membership functions for the one-dimensional case.

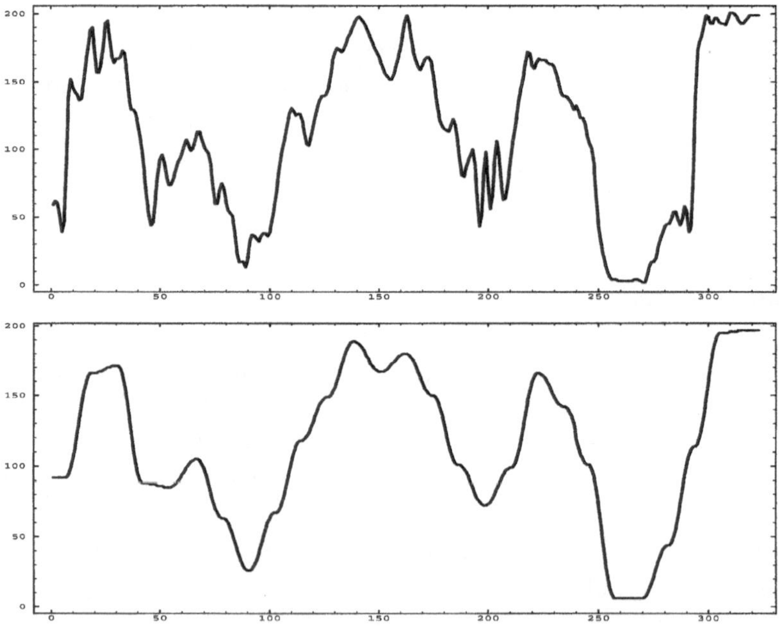


Figure 5.3 An example of compression/decompression with the membership functions of Figure 5.2. The original signal is on top, the signal obtained after decompression is at the bottom.

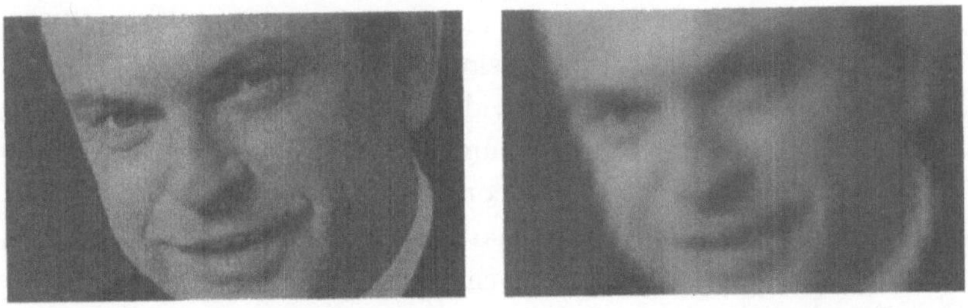

Figure 5.4 The original picture (left) and the decompressed picture (right).

When the form of compression and decompression membership functions is completely unrestricted, convergence of the evolutionary algorithm is rather slow, in the order of the hundreds of generations.

To achieve better performance, the assumption has been made that any compression or decompression membership function has to be symmetrical with respect to the center of a square. Finding a shape for such a membership function boils down to determining its shape along the radius departing from the center of the square. This device dramatically decreases the number of degrees of freedom and, therefore, it speeds up convergence of the algorithm.

## 5.6 A Case Study: Breast Cancer Diagnosis

We finish this chapter by presenting a fuzzy-evolutionary approach to breast cancer diagnosis developed by Carlos Andrés Peña-Reyes and Moshe Sipper. The work is described in full detail in [168].

### 5.6.1 Motivation

A major class of problems in medical science involves the diagnosis of diseases, based upon various tests performed upon the patient. When several tests are involved, the ultimate diagnosis may be difficult to obtain, even for a medical expert. This has given rise, over the past few decades, to computerized diagnostic tools, intended to aid the physician in making sense out of the welter of data. A prime target for such computerized tools is in the domain of cancer diagnosis.

A good computerized diagnostic tool should possess two characteristics, which are often in conflict. First, the tool must attain the highest possible *performance*, i.e., diagnose the presented cases correctly as being either *benign* or *malignant*. Moreover, it would be highly desirable to be in possession of a so-called *degree of confidence*: the system not only provides a binary diagnosis (benign or malignant), but also outputs a numeric value which represents the degree to which the system is confident about its response. Second, it would be highly beneficial for such a diagnostic system to be human-friendly, exhibiting so-called *interpretability*. This means that the physician is not faced with a black box that simply spouts answers (albeit correct) with no explanation; rather, we would like the system to provide some insight into *how* it derives its outputs.

As we have seen in Chapter 3, fuzzy systems do favor interpretability. However, finding good fuzzy systems can be quite an arduous task. This is where evolutionary algorithms step in, enabling the automatic production of fuzzy systems, based on a data set of training cases.

*Fuzzy modeling* is the task of identifying the parameters of a fuzzy inference system so that a desired behavior is attained [249]. With the *direct* approach, a fuzzy model is constructed using knowledge from a human expert. This task becomes difficult when the available knowledge is incomplete or when the problem space is very large, thus motivating the use of *automatic* approaches to fuzzy modeling.

The parameters of fuzzy inference systems can be classified into four categories: logical, structural, connective, and operational. Generally speaking, this order also represents their relative influence on performance, from most influential (logical) to least influential (operational).

In fuzzy modeling, logical parameters, like the inference mechanism, the choice of fuzzy logical connectors, and the defuzzification method, are usually predefined by the designer based on experience and on problem characteristics.

### 5.6.2 The Wisconsin Breast Cancer Diagnosis Problem

Breast cancer is the most common cancer among women, excluding skin cancer. The presence of a breast mass is an alert sign, but it does not always indicate a malignant cancer. Fine needle aspiration (FNA) of breast masses, an outpatient procedure that uses a small-gauge needle to extract fluid directly from a breast mass is a cost-effective, non-traumatic, and mostly non-invasive diagnostic test that obtains information needed to evaluate malignancy.

The Wisconsin breast cancer diagnosis (WBCD) data set [142] is the result of the efforts of experts at the University of Wisconsin Hospital for accurately diagnosing breast masses based solely on an FNA test [134]. Nine visually assessed characteristics of an FNA sample considered relevant for diagnosis were identified, and assigned an integer value between 1 and 10.

The data set itself consists of 683 cases, with each record representing the classification for a certain ensemble of measured values:

| *case* | $v_1$ | $v_2$ | $v_3$ | $\cdots$ | $v_9$ | *diagnostic* |
|---|---|---|---|---|---|---|
| 1 | 5 | 1 | 1 | $\cdots$ | 1 | *benign* |
| 2 | 5 | 4 | 4 | $\cdots$ | 1 | *benign* |
| $\vdots$ | $\vdots$ | $\vdots$ | $\vdots$ | $\ddots$ | $\vdots$ | $\vdots$ |
| 683 | 4 | 8 | 8 | $\cdots$ | 1 | *malignant* |

Note that the diagnostics do not provide any information about the degree of benignity or malignancy.

There are several studies based on this data set. Bennet and Mangasarian [22] used linear programming techniques, obtaining a 99.6% classification rate on 487 cases (the reduced database available in 1992). However, their solution exhibits little explanatory power, i.e., diagnostic decisions are essentially black boxes, with no explanation as to how they were attained. With increased interpretability in mind as a prior objective, a number of researchers have applied the method of extracting Boolean rules from neural networks [205, 206, 215]. Their results are encouraging, exhibiting both good performance and a reduced number of rules and relevant input variables. Nevertheless, these systems use Boolean rules and are not capable of furnishing the user with a measure of confidence for the decision made.

### 5.6.3 Evolving Fuzzy Systems for the WBCD Problem

The solution scheme proposed for the WBCD problem is depicted in Figure 5.5. It consists of a fuzzy system and a threshold unit. The fuzzy system computes a continuous appraisal value of the malignancy of a case, based on the input values. The threshold unit then outputs a *benign* or *malignant* diagnostic according to the fuzzy system's output.

In order to evolve the fuzzy model, one must make some preliminary decisions about the fuzzy system itself and about the genetic algorithm encoding.

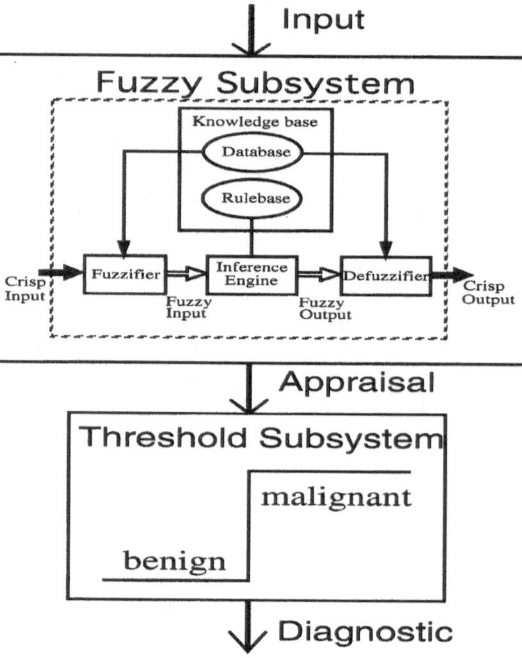


Figure 5.5 The proposed diagnosis system.

#### Fuzzy System Parameters

Previous knowledge about the WBCD problem and about some of the extant rule-based models represents valuable information to be used for the choice of fuzzy parameters. The following results, described in previous works, were taken into consideration:

- *Small number of rules.* Systems with no more than four rules have been shown to obtain high performance [167, 205].
- *Small number of variables.* Rules with no more than four antecedents have proven adequate [167, 206, 215].
- *Monotonicity of the input variables.* Simple observation of the input and output spaces shows that higher-valued variables are associated with malignancy.

Some fuzzy models forgo interpretability in the interest of improved performance. Where medical diagnosis is concerned, interpretability—also called linguistic integrity—is the major advantage of fuzzy systems. Thus, the following semantic criteria defining constraints on the fuzzy parameters were taken into account [172]:

- *Distinguishability.* Each linguistic label should have semantic meaning and its associated fuzzy set should clearly define a

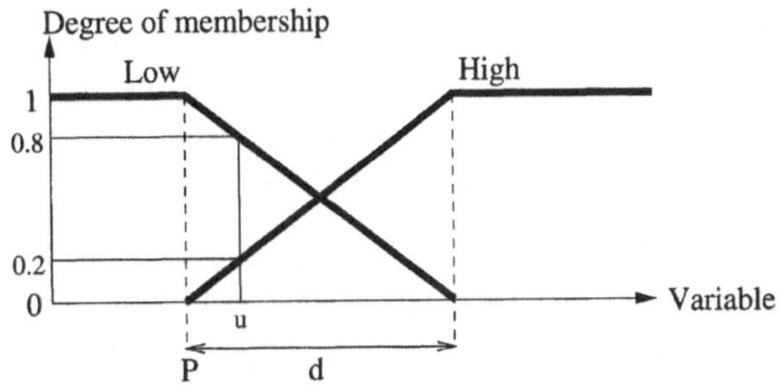


Figure 5.6 Input variable with two possible linguistic values labeled *Low* and *High*, and orthogonal membership functions, plotted above as degree of membership versus input values. $P$ and $d$ define the start point and the length of membership function edges, respectively. The orthogonality condition means that the sum of all membership functions at any point is one. In the figure, an example value $u$ is assigned the membership values $\mu_{Low}(u) = 0.8$ and $\mu_{High}(u) = 0.2$ (as it can be seen, $\mu_{Low}(u) + \mu_{High}(u) = 1$).

range in the universe of discourse. For example, to describe variable *Clump Thickness* ($v_1$) one may use two labels: *Thick* and *Diffuse* (but the authors opted for *Low* and *High* for uniformity with other variables). Their membership functions are defined using parameters $P$ and $d$, as shown in Figure 5.6.

- *Justifiable number of domains.* The number of membership functions of a variable should be compatible with the number of conceptual entities a human being can handle. This number should not exceed the limit of $7 \pm 2$ distinct terms. The same criterion is applied to the number of variables in the rule antecedent. For example, the following would be considered an adequate rule:

  IF $v_1$ is *High* AND $v_2$ is *High* AND $v_4$ is *High* AND $v_6$ is *Low* AND $v_8$ is *Low*
  THEN *output* is *benign*.

- *Coverage.* Any element from the universe of discourse should belong to at least one of the fuzzy sets. That is, its membership value must be nonzero for at least one of the linguistic labels. Referring to Figure 5.6, we see that any value along the $x$-axis belongs to at least one fuzzy set (*Low*, *High*, or both); no value lies outside the range of all sets.

- *Normalization.* Since all labels have semantic meaning, then, for each label, at least one element of the universe of discourse should have a membership value equal to one. In Figure 5.6, we observe that both *Low* and *High* are normal fuzzy sets.
- *Orthogonality.* For each element of the universe of discourse, the sum of all its membership values should be equal to one (e.g., in Figure 5.6 *Low* membership value of 0.8 entails a *High* membership value of 0.2).

The resulting setup for the fuzzy systems is the following:

1. *Logical parameters*
   - Inference mechanism: Sugeno-type fuzzy system, with constant consequents.
   - Fuzzy operators: min and max.
   - Input membership function type: orthogonal, trapezoidal (see Figure 5.6).
   - Defuzzification method: weighted average.
2. *Structural parameters*
   - Relevant variables: given insufficient *a priori* knowledge to define them, their selection was left to the evolutionary algorithm.
   - Number of input membership functions: two membership functions, denoted *Low* and *High* (Figure 5.6).
   - Number of output membership functions: two singletons are used, corresponding to the *benign* and *malignant* diagnostics.
   - Number of rules: this is a user-configurable parameter. Based on previous results [167], the number of rules was limited to be between one and five. The rules themselves were to be found by the genetic algorithm.
3. *Connective parameters*
   - Antecedents of rules: to be found by the genetic algorithm.
   - Consequent of rules: the algorithm finds rules for the *benign* diagnostic; the *malignant* diagnostic is a default ELSE condition.

- Rule weights: active rules have a weight of value 1, and the `else` condition has a weight of 0.25.

4. *Operational parameters*

   - Input membership function values: to be found by the genetic algorithm.
   - Output membership function values: following the WBCD data set, a value of 2 was used for *benign* and 4 for *malignant*.

**The Evolutionary Algorithm**

Three aspects of the fuzzy system are evolved: the choice of relevant variables, their input membership function values, and the antecedents of rules.

The genome, in the form of a fixed-length bit string, is constructed as follows:

- Membership function parameters. There are nine input variables ($v_1$ to $v_9$), each with two parameters $P$ and $d$, defining the start point and the length of the membership function edges, respectively (see Figure 5.6).
- Rule antecedents. The $i$-th rule has the form:

  IF $v_1$ is $A_1^i$ AND ... AND $v_9$ is $A_9^i$
  THEN *output* is *benign*,

  where $A_j^i$ represents the membership function applicable to variable $v_j$. $A_j^i$ can take on the values: 1 (*Low*), 2 (*High*), or 0 or 3 (*Any*).
- Relevant variables are searched for implicitly by letting the algorithm choose a membership function whose value is one over the whole domain (the *Any* linguistic label) as valid antecedents; in such case, the actual value of the variable becomes irrelevant.

Table 5.1 delineates the parameters encoding, which together form a single individual's genome. Figure 5.7 shows a sample genome.

To evolve the fuzzy system, a genetic algorithm with a fixed population size of 200 individuals and fitness-proportionate selection was used. The algorithm terminates when the maximum number

| Parameter | Values | Bits | Quantity | Total bits |
|---|---|---|---|---|
| $P$ | [1-8] | 3 | 9 | 27 |
| $d$ | [1-8] | 3 | 9 | 27 |
| $A$ | [0-3] | 2 | $9N_r$ | $18N_r$ |

Table 5.1 Parameters encoding of an individual's genome. Total genome length is $54 + 18N_r$, where $N_r$ denotes the number of rules ($N_r$ is set *a priori* to a value between 1–5, and remains fixed during the run).

of generations, $G_{max}$ is reached or when the increase in fitness of the best individual over five successive generations falls below a certain threshold.

The fitness function combines three criteria:

1. $F_c$, the classification performance, computed as the percentage of cases correctly classified;

2. $F_e$, the squared difference between the continuous appraisal value (in the range $[2, 4]$) and the correct discrete diagnosis given by the WBCD database (either 2 or 4);

3. $F_v$, the average number of variables per active rule.

The fitness function is given by

$$F = F_c - \alpha F_v - \beta F_e,$$

where $\alpha = 0.05$ and $\beta = 0.01$ (these latter values were derived empirically). $F_c$, the ratio of correctly diagnosed cases, is the most important measure of performance. $F_v$ measures the linguistic integrity (interpretability), penalizing systems with a large number of variables per rule (on average). Finally, $F_e$ adds selection pressure towards systems with low squared error.

### 5.6.4 Experiments and Results

The evolutionary experiments performed fall into three categories, depending on how the available data were repartitioned into two distinct sets: training set and test (or evaluation) set.

The three experimental settings are:

1. the training set contains all 683 cases of the WBCD database, while the test set is empty;

2. the training set contains 75% of the WBCD cases, and the test set contains the remaining 25% of the cases;

| $P_1$ | $d_1$ | $P_2$ | $d_2$ | $P_3$ | $d_3$ | $P_4$ | $d_4$ | $P_5$ | $d_5$ | $P_6$ | $d_6$ | $P_7$ | $d_7$ | ... |
|---|---|---|---|---|---|---|---|---|---|---|---|---|---|---|
| 4 | 3 | 1 | 5 | 2 | 7 | 1 | 1 | 1 | 6 | 3 | 7 | 4 | 6 | ... |

| ... | $A_1^1$ | $A_2^1$ | $A_3^1$ | $A_4^1$ | $A_5^1$ | $A_6^1$ | $A_7^1$ | $A_8^1$ | $A_9^1$ |
|---|---|---|---|---|---|---|---|---|---|
| ... | 0 | 1 | 3 | 3 | 2 | 3 | 1 | 3 | 1 |

(a)

**Database**

| | $v_1$ | $v_2$ | $v_3$ | $v_4$ | $v_5$ | $v_6$ | $v_7$ |
|---|---|---|---|---|---|---|---|
| $P$ | 4 | 1 | 2 | 1 | 1 | 3 | 4 |
| $d$ | 3 | 5 | 7 | 1 | 6 | 7 | 6 |

**Rule base**

Rule 1 IF $v_2$ is *Low* AND $v_5$ is *High* AND $v_7$ is *Low* THEN *output* is *benign*

Default ELSE *output* is *malignant*

(b)

Figure 5.7 Example of a genome for a single-rule system. (a) Genome encoding. The first 18 positions encode the parameters $P$ and $d$ for the nine variables $v_1$–$v_9$ (only seven are shown). The rest encode the membership function applicable for the nine antecedents of each rule. (b) Interpretation. Database and rule base of the single-rule system encoded by (a). The parameters $P$ and $d$ are interpreted as illustrated in Figure 5.6.

3. the training set contains 50% of the WBCD cases, and the test set contains the remaining 50% of the cases.

In the last two categories, the choice of training-set cases is random, and performed anew at the outset of every evolutionary run. The number of rules per system was also fixed at the outset, to be between one and five, i.e., evolution seeks a system with an *a priori* given number of rules (the choice of number of rules per system determines the final structure of the genome, as presented in Table 5.1).

A total of 120 evolutionary runs were performed, all of which found systems whose classification performance exceeds 94.5%. In particular, considering the best individual per run (i.e., the evolved system with the highest classification success rate), 78 runs led to a fuzzy system whose performance exceeded 96.5%, and of these, 8 runs found systems whose performance exceeded 97.5%.

When compared with other rule-based diagnostic approaches, the results obtained by this method score very well. Indeed, the evolved fuzzy systems surpass those obtained by previous approaches in terms of both performance and simplicity of rules. Not only is high performance exhibited, but, moreover, the combined fuzzy-

Rule 1
IF $v_3$ is *Low* AND $v_7$ is *Low* AND $v_8$ is *Low* AND $v_9$ is *Low*
THEN *output* is *benign*
Rule 2
IF $v_1$ is *Low* AND $v_2$ is *Low* AND $v_3$ is *High* AND $v_4$ is *Low* AND $v_5$ is *High* AND $v_9$ is *Low*
THEN *output* is *benign*
Rule 3
IF $v_1$ is *Low* AND $v_4$ is *Low* AND $v_6$ is *Low* AND $v_8$ is *Low*
THEN *output* is *benign*
Default
ELSE *output* is *malignant*

Figure 5.8 The best evolved, fuzzy diagnostic system with three rules. It exhibits an overall classification rate of 97.8%, and an average of 4.7 variables per rule.

evolutionary approach enables the introduction of a confidence measure of the diagnostic decision (see Subsection 5.6.4).

#### Discussion of Results

Here two of the top-performance evolved systems are described. The first system, shown in Figure 5.8, consists of three rules (the default else condition is not counted as an active rule). Taking into account all three criteria of performance (classification rate, number of rules per system, and average number of variables per rule) this system can be considered the top one over all 120 evolutionary runs. It obtains 98.7% correct classification rate over the benign cases, 97.07% correct classification rate over the malignant cases,[1] and an overall classification rate (i.e., over the entire database) of 97.8%.

A thorough test of this three-rule system revealed that the second rule (Figure 5.8) is never actually used. Thus, it can be eliminated altogether from the rule base, resulting in a two-rule system (also reducing the average number of variables per rule from 4.7 to 4).

Can the genetic algorithm automatically discover a two-rule system, i.e., without recourse to any post-processing (such as that described in the previous paragraph)? The results have shown that this is indeed the case—one such solution is presented in Figure 5.9. It obtains a correct classification rate of 97.3% over the benign cases, of 97.49% over the malignant cases, and an overall classification rate of 97.36%.

[1]The WBCD data set contains 444 benign cases and 239 malignant cases.

Rule 1
IF $v_1$ is *Low* AND $v_3$ is *Low*
THEN *output* is *benign*
Rule 2
IF $v_2$ is *Low* AND $v_5$ is *Low* AND $v_6$ is *Low* AND $v_8$ is *Low*
THEN *output* is *benign*
Default
ELSE *output* is *malignant*

Figure 5.9 The best evolved, fuzzy diagnostic system with two rules. It exhibits an overall classification rate of 97.36%, and an average of three variables per rule.

**Diagnostic confidence**

Up until now, the evolved fuzzy systems have been used to ultimately produce a binary classification value, either *benign* or *malignant*, with no finer gradations.

Going back to Figure 5.5, it can be seen that the diagnostic system comprises in fact two subsystems: the first subsystem consists of the (evolved) fuzzy system, which, upon presentation of an input (in this case, a WBCD database entry) proceeds to produce a *continuous* appraisal value; this value is then passed along to the second subsystem (the threshold unit) which produces the final binary output (*benign* or *malignant*). The first subsystem (the fuzzy system) is the one actually evolved. The threshold subsystem simply outputs *malignant* if the appraisal value is below a fixed threshold value, and outputs *benign* otherwise. The threshold value is assigned by the user through knowledge of the problem at hand.

As explained in Chapter 3, each rule gives a contribution to the overall output of the system that is proportional to the degree of truth of its antecedent. The appraisal value computed by the evolved fuzzy systems is in the range [2, 4]. It seems a reasonable choice to place the threshold value at 3 (i.e., in the middle of the range), with inferior values classified as benign, and superior values classified as malignant.

Therefore, by looking at how close the appraisal value produced by the system is to an extreme of the range [2, 4], an estimate of the confidence in the response given can be obtained. This demonstrates a prime advantage of fuzzy systems, namely, the ability to output not only a (crisp) classification, but also a measure representing the system's confidence in its output. The three-rule system of Figure 5.8 computes intermediate appraisal values (between, say, 2.4 and 3.6) for 39 cases; these might thus be considered the cases for which we are somewhat less confident about the output.

Chapter 6

# Neuro-fuzzy Systems

## 6.1 Introduction

THIS CHAPTER deals with neuro-fuzzy systems, i.e., those soft computing methods that combine in various ways neural networks and fuzzy concepts. Each methodology has its particular strengths and weaknesses that make it more or less suitable in a given context. For example, fuzzy systems can reason with imprecise information and have good explanatory power. On the other hand, rules for fuzzy inference have to be explicitly built into the system or communicated to it in some way; in other words the system cannot learn them automatically. Neural networks represent knowledge implicitly, are endowed with learning capabilities, and are excellent pattern recognizers. But they are also notoriously difficult to analyze: to explain how exactly they reach their conclusions is far from easy while the knowledge is explicitly represented through rules in fuzzy systems.

The complementarity between fuzzy systems and learning systems, especially ANNs, has been recognized early by researchers. Taking again a rather courageous, and utterly unrealistic biological analogy, we could say that, conceptually at least, mixed neural and fuzzy systems resemble nervous systems where neural cells are the low-level perceptive and signal integration part that make possible the higher level functions of the brain such as reasoning and linguistic abilities. In this metaphor the ANN part stands for the perceptive and signal processing biological machinery, while the fuzzy part represents the emergent "higher level" reasoning aspects. As a result, these two technologies have been integrated in various ways, giving rise to *hybrid* systems that are able to overcome many of the limitations of the individual techniques. Therefore, neuro-fuzzy systems are likely to be of wider applicability on real-life problems. The reader should be aware that the field has an enormous variety and it would be impossible to present a complete survey in a single chapter. We have thus been obliged to make a choice of topics that we believe are significant and representative of modern trends but by no means exhaustive. Two useful recent books covering in detail all the topics treated here and more are Nauck *et al.* [156] and Jang *et al.* [104].

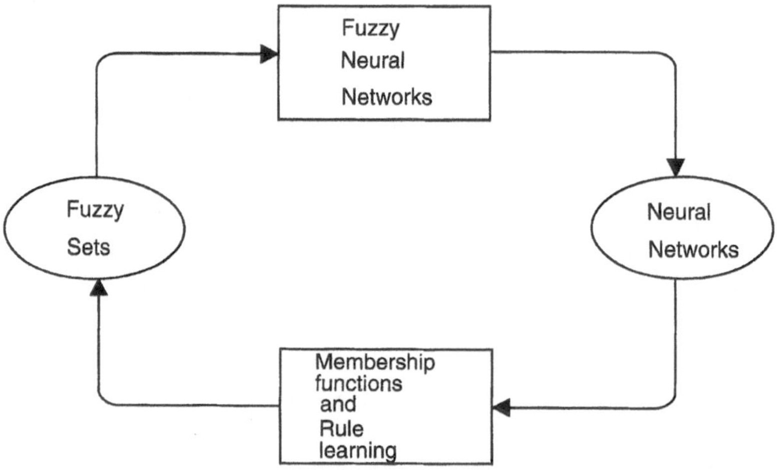


Figure 6.1 Schematic view of how artificial neural networks and fuzzy systems can interact synergetically.

There are two main ways in which ANNs and fuzzy systems can interact synergetically. One is the "fuzzification" of neural networks and the other consists in endowing fuzzy system with neural learning features. In the first case, fuzziness may be introduced at different levels in a neural net: the weight level, the transfer function level, or the learning algorithm level. In the second case, the most common arrangement is for the ANN to learn membership functions or rules for a given fuzzy system. This relationships are schematically depicted in Figure 6.1.

More precisely, according to [156], systems that combine neural and fuzzy ideas can be divided into three classes:

- *co-operative*, in the sense that neural algorithms adapt fuzzy systems, which can in turn be distinguished in
  - off-line (neural algorithms learn membership functions or rules or both, once and for all)
  - on-line (neural algorithms are used to adapt the membership functions or the rules of the fuzzy system or both, as the system operates);
- *concurrent* (but we would prefer to call them *sequential*), where the two techniques are applied after one another as pre- or post-processing.
- *hybrid* (here too, this terminology can be misleading, because, as a matter of fact, all neuro-fuzzy systems are hybrid), the fuzzy system being represented as a network structure, making it possible to take advantage of learning algorithms inherited from ANNs. From now on this combination will be called "fuzzy neural networks".

Concurrent (i.e., sequential) approaches are the weakest form of combination between neural and fuzzy techniques, and not such an interesting one for the purpose of this chapter. After all, the two techniques retain their individualities and can be understood without studying their interactions. Therefore, the first part of the chapter describes "fuzzy neural networks", that is, how single neural units and networks can be given a fuzzy flavor. The second part of the chapter deals with the other aspect of the mutual relationship that is, using ANNs to help design efficient fuzzy systems "cooperatively".

## 6.2 Fuzzy Neural Networks

The purpose of this section is to introduce fuzzy concepts into single artificial neurons and neural networks. Fuzzy systems and neural networks are certainly different soft computing paradigms; however, they are rather complementary if one takes into account their respective strong and weak features. Therefore, integrating them into a single new soft computing model gives hopes of exploiting their complementary nature by reinforcing the good points and by alleviating their respective shortcomings. Fuzziness can be introduced in several ways into artificial neurons. In the next section we present a way of "fuzzifying" single neurons.

### 6.2.1 Fuzzy Neurons

Fuzzy models of artificial neurons can be constructed by using fuzzy operations at the single neuron level. The fuzzy operations that have been used for that purpose are the union and the intersection of fuzzy sets and, more generally, *t-norms* and *t-conorms* also called *s-norms* which are extensions of the usual fuzzy operations (see Chapter 3, Section 3.2.4). A variety of fuzzy neurons can be obtained by applying fuzzy operations to connection weights, to aggregation functions or to both of them. We shall start from the structure of an artificial neuron such as it was introduced in Chapter 2 and which is reproduced here for ease of reference in Figure 6.2.

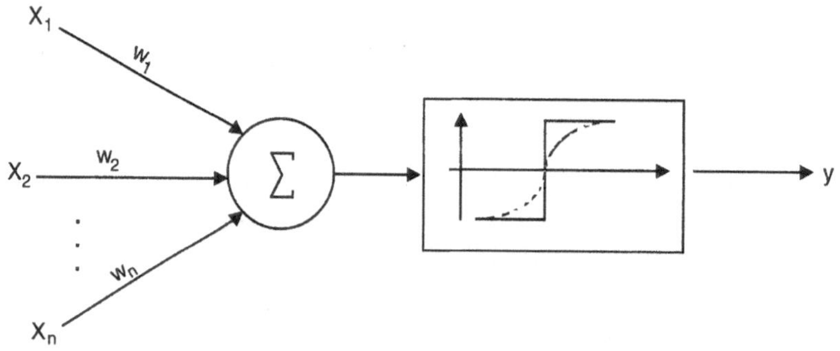


Figure 6.2 Model of a crisp artificial neuron.

We recall that a vector $\mathbf{x} = (x_1, x_2, \ldots, x_n)$ of input values enters a neural unit through $n$ incoming connections after having being modified by the corresponding connection weights $\mathbf{w} = (w_1, w_2, \ldots, , w_n)$. The neuron computes the sum of the weighted inputs, which is simply the scalar product $\mathbf{w} \cdot \mathbf{x}$ and produces an output signal according to a predefined activation function $g$. When the function is a simple step function the neuron fires (i.e, it produces a 1 output signal) if $\mathbf{w} \cdot \mathbf{x}$ reaches a given thereshold value, otherwise it doesn't fire (the output is 0). However, for numerical reasons, it is often useful to have $g$ as a non-linear monotonic mapping of the type:

$$g\colon [0,1] \to [0,1], \tag{6.1}$$

such as a sigmoid, hyperbolic tangent or gaussian curve. In this case the neuron emits a graded output signal between 0 and 1 (See Chapter 2, Equation 2.2 and Equation 2.3).

The above standard model of an artificial neuron derives some credibility from biological data on neural cells but it is certainly a gross oversimplification of reality. Thus, although it is a convenient choice, there is nothing special about the weighted sum of the input values as an aggregation operator. A step toward the fuzzification of an artificial neuron can be done by considering other forms $A$ of the aggregation function according to the more general equation:

$$y = g\left(A(\mathbf{w}, \mathbf{x})\right), \tag{6.2}$$

where $g$ is the transfer function and $y$ is the scalar output signal of the neuron.

In fact, fuzzy union, fuzzy intersection and, more generally, *s-norms* and *t-norms* can be used as an aggregation function for the weighted inputs to an artificial neuron. Due to the fact that triangular norms form an infinite family, there exist an infinite number of possibilities for defining fuzzy neurons at the level of the aggregation function. In what follows, we present a particular class of fuzzy neurons, the OR and AND neurons.

### OR Fuzzy Neuron

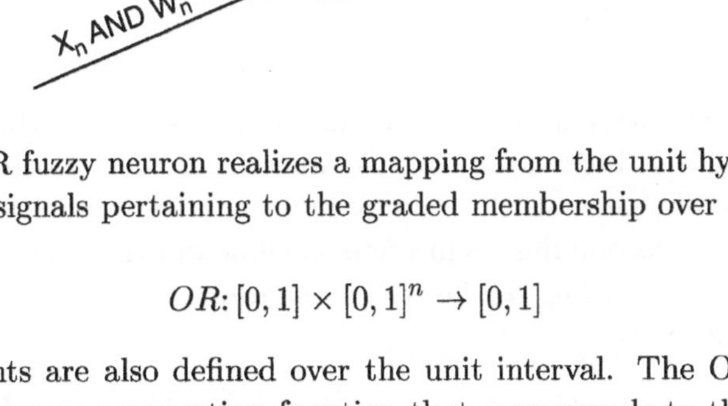

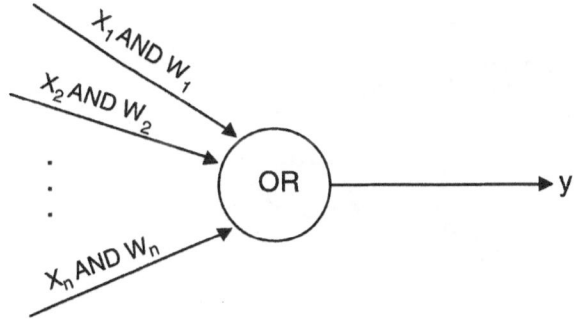


Figure 6.3 Model of an OR fuzzy neuron.

The OR fuzzy neuron realizes a mapping from the unit hypercube to fuzzy signals pertaining to the graded membership over the unit interval:

$$OR\colon [0,1] \times [0,1]^n \to [0,1]$$

the weights are also defined over the unit interval. The OR fuzzy neuron uses an aggregation function that corresponds to the maximum of the weighted inputs; that is, it selects the fuzzy disjunction of the weighted inputs as follows:

$$y = OR(x_1\ AND\ w_1, x_2\ AND\ w_2, \ldots, x_n\ AND\ w_n). \qquad (6.3)$$

This setting is depicted in Figure 6.3. As we saw in Chapter 3, Section 3.2.4, fuzzy set logical connectives are usually defined in terms of triangular norms $\triangle$ and t-conorms $\triangledown$. Thus, the preceding expression for the neuron output becomes:

$$y = \triangledown_{j=1}^{n}(x_j \ \triangle\ w_j). \qquad (6.4)$$

The transfer function $g$ is linear. In a manner analogous to standard ANNs a bias term can be added representing a constant $x_0 = 0$ value input signal with a weight $w_0$. Taking the bias into account the preceding equation reads:

$$y = \triangledown_{j=0}^{n}(x_j \ \triangle\ w_j). \qquad (6.5)$$

We observe that, for any connection $k$, if $w_k = 0$ then $w_k\ AND\ x_k = 0$ while if $w_k = 1$ then $w_k\ AND\ x_k = x_k$ independent of $x_k$.

### AND Fuzzy Neuron

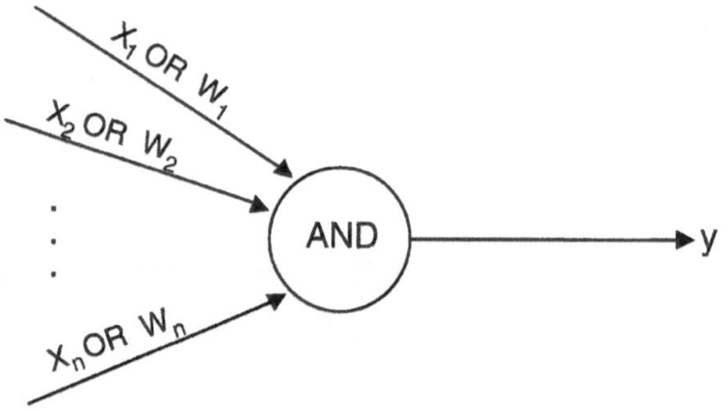


Figure 6.4 Model of an AND fuzzy neuron.

The AND fuzzy neuron (see Figure 6.4) is similar to the OR case except that it takes the fuzzy conjunction of the weighted inputs, which is the same as the minimum. First the inputs are "ored" with the corresponding connection weights and then the results are aggregated according to the AND operation. The transfer function $g$ is again linear:

$$y = AND(x_1\ OR\ w_1, x_2\ OR\ w_2, \ldots, x_n\ OR\ w_n). \qquad (6.6)$$

Analogous to the OR case, when expressed in terms of triangular norms the output fuzzy value is given by:

$$y = \Delta_{j=0}^{n}(x_j \;\nabla\; w_j), \tag{6.7}$$

where the bias term $x_0$ is now equal to 1, giving $w_0 \; AND \; x_0 = w_0$.

Of course, in the generalized forms based on t-norms, operators other than *min* and *max* can be used such as algebraic and bounded products and sums. As they stand, both the OR and the AND logic neurons are *excitatory* in character, that is higher values of $x_k$ produce higher values of the output signal $y$. The issue of *inhibitory* (negative) weights deserves a short digression since their introduction is not as straightforward as it is in the standard neural networks. In fact, we are here in the realm of fuzzy sets and we would obviously like to maintain the full generality of the operations defined in $[0, 1]$. If the interval is extended to $[-1, 1]$ as it is customary in ANNs, logical problems arise from the fuzzy set-theoretical point of view. The proper solution to make a weighted input inhibitory is to take the fuzzy complement of the excitatory membership value $\bar{x} = 1 - x$. In this way, the generic input vector $\mathbf{x} = (x_1, x_2 \ldots x_n)$ now includes the complemented values as well: $\mathbf{x} = (x_1, x_2 \ldots x_n, \bar{x}_1, \bar{x}_2 \ldots \bar{x}_n)$. The weighted inputs $x_i \;\circ\; w_i$, where $\circ$ is a t-norm or t-conorm, can be general fuzzy relations too, not just simple products as in standard neurons. As well, the transfer function $g$, which has been supposed linear, can be a non-linear one such as a sigmoid. These kinds of fuzzy neurons have been introduced by Pedrycz and coworkers [173, 195].

### OR/AND Fuzzy Neuron

A generalization of the above simple fuzzy neurons is the OR/AND neuron [95]. The OR/AND neuron is a combination of the AND and OR neurons into a two-layer structure as depicted in Figure 6.5. Taken as a whole, this structure can produce a spectrum of intermediate behaviors that can be modified in order to suit a given problem. Looking at the figure, it is apparent that the behavior of the net can be modulated by suitably weighting the output signals from the OR or the AND parts through setting or learning the connection weights $c_1$ and $c_2$. The limiting cases are $c_1 = 0$ and $c_2 = 1$ where the system reduces itself to a pure AND neuron and the converse $c_1 = 1$, $c_2 = 0$, in which case the behavior corresponds to that of a pure OR neuron.

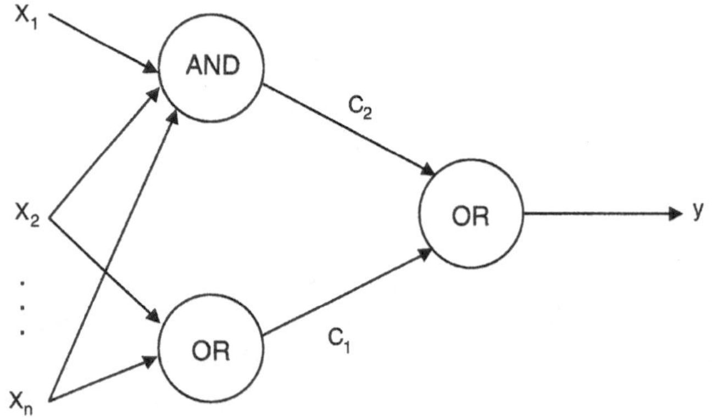


Figure 6.5 The computational structure of an AND/OR fuzzy neuron.

### 6.2.2 Multilayered Fuzzy Neural Networks

Fuzzy neurons such as those described in the previous section can be assembled together into multilayered networks [170]. Since the neurons used to build the nets are in general different, the construction gives rise to non-homogeneous neural networks, in contrast with the usually homogeneous networks that are used in the crisp ANN domain. For example, Figure 6.6 dcpicts a two-layer network (not counting the input layer) composed of a first layer with $p$ neurons of the same AND type and a second output layer wich aggregates all the preceding signals with a single OR neuron. The input is constituted by $2n$ values including both the direct and the complemented ones. A second possibility is to have OR neurons in the hidden layer and a single AND neuron in the output layer (Figure 6.7). These two types of networks have been called *logic processors* by Pedrycz.

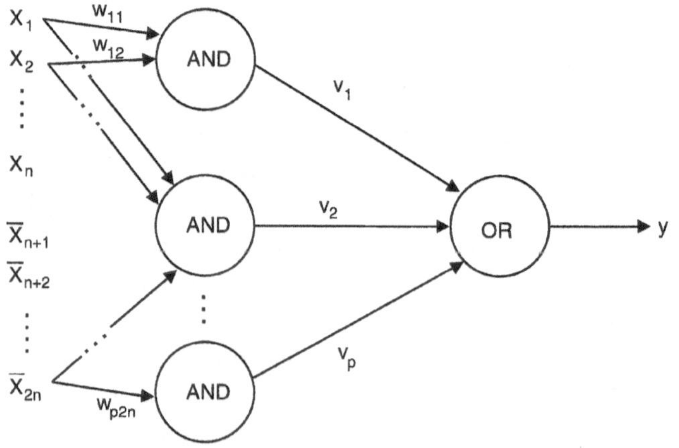


Figure 6.6 A three-layer artificial neural network with all AND fuzzy units in the hidden layer.

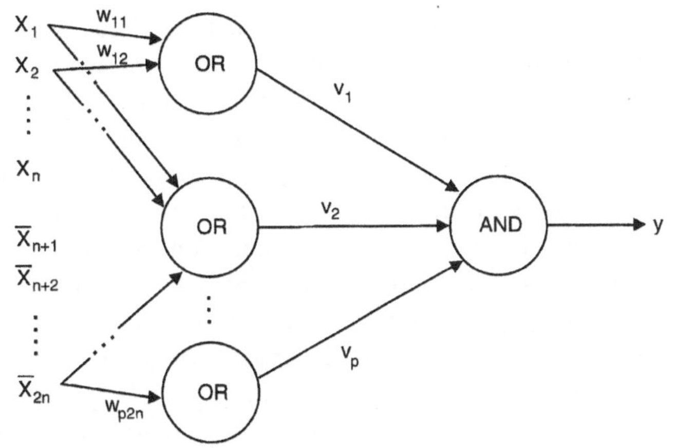


Figure 6.7 A three-layer artificial neural network with all OR fuzzy units in the hidden layer.

Following Pedrycz [170], for the first layered network type we have that the signals $z_k$ produced at the hidden layer are given by:

$$z_k = [\triangle_{j=1}^{n}(x_j \ \triangledown \ w_{k,j})] \ \triangle \ [\triangle_{j=1}^{n}(\bar{x}_j \ \triangledown \ w_{k,(n+j)})], \quad k = 1, 2, \ldots, p; \tag{6.8}$$

where $\mathbf{w}_k$ is the vector of all connection weights from the inputs to the $k$th node of the hidden layer. The output value $y$ is the single OR-aggregation of the previous signals from the hidden layer:

$$y = \triangledown_{j=1}^{p}(x_j \ \triangle \ v_j). \tag{6.9}$$

In the last expression the vector $\mathbf{v}$ is the weight vector of all the connections from the hidden layer nodes to the single output node. The network in which all the hidden nodes are of the OR type and the output node is an AND fuzzy neuron (Figure 6.7) gives analogous expressions with the $\triangle$ and $\triangledown$ symbols exchanged. If we restrict ourselves to the pure two-valued Boolean case then the network with the hidden OR layer represents an arbitrary Boolean function as a sum of minterms, while the second network is its dual in the sense that it represents any Boolean function as a product of maxterms. More generally, if the values are continuous members of a fuzzy set then these networks approximate a certain unknown fuzzy function.

### 6.2.3 Learning in Fuzzy Neural Networks

The interest of having fuzzy connectives organized in network form is that there are thus several ways in which ANN supervised learning methods can be applied to the fuzzy structure. This is a definite plus in many situations since learning capabilities are not typical of fuzzy systems. Of course, there exist other ways in which fuzzy systems can learn but this particular neuro-fuzzy hybrid is useful in light of the large amount of knowledge that has been accumulated on the crisp ANNs versions. Supervised learning in fuzzy neural networks consists in modifying their connection weights in a such a manner that an error measure is progressively reduced by using sets of known input/output data pairs. Another important requirement is that the network thus obtained be capable of generalization; that is, its performance should remain acceptable when it is presented with new data (see the discussion on ANN supervised learning in Chapter 2, Section 2.5).

Let us call the set of $n$ training data pairs $(\mathbf{x}_k, d_k)$ for $k = 1, 2 \ldots n$, where $\mathbf{x}_k$ is a vector of input data and $d_k$ is the corresponding observed scalar output. A single fuzzy neuron adapts its connection weights in order to reduce a measure of error averaged over the training set:

$$\mathbf{w}^{t+1} = \mathbf{w}^t + \Delta \mathbf{w}^t, \tag{6.10}$$

where the weight change is a given function $F$ of the difference between the target response $d$ and the calculated node output $y$:

$$\Delta \mathbf{w}^t = F(|d^t - y^t|), \tag{6.11}$$

For instance, in standard ANNs a common learning rule is the *delta* rule, which uses an estimate of the gradient of the continuous neuron activation to reduce the mean square error (Chapter 2, Section 2.5). For fuzzy neurons, one should take into account the fact that the weighted inputs are not simple scalar products; rather, they are more complex relationships between fuzzy sets.

For the whole network supervised learning proceeds as follows. A criterion function $E$ is defined such that it gives a mesure of how well the fuzzy network maps input data into the corresponding output. A common form for $E$ is the sum of the squared errors:

$$E(\mathbf{w}) = 1/2 \sum_{k=1}^{n} (d_k - y_k)^2 \tag{6.12}$$

The goal of the learning algorithms is to systematically vary the connection weights in such a way that $E$ is minimized on the training data set. This can be achieved by taking the gradient descent of $E$ with respect to the weights. The update step in the output weights $w_{i,j}$ of the connections between unit $j$ in the hidden layer and unit $j$ in the output layer can thus be found by differentiating:

$$\Delta w_{i,j} = -\eta \frac{\partial E}{\partial w_{i,j}}, \tag{6.13}$$

where $\eta$ is a scale factor that controls the magnitude of the change. The update in the input weights can be found by using the chain rule for partial derivatives in the backpropagation style (see also Chapter 2, Section 2.5). The derivation of the weight changes layer by layer back to the input layer is straightforward but somewhat tricky. The interested reader can find an example completely worked out in Pedrycz's book [170]. The network chosen for the example corresponds to the three-layer system of Equation 6.8 and Equation 6.9 where the algebraic sum is used for the s-norm and the product for the t-norm.

### 6.2.4 An Example: NEFPROX

Approximating a continuous unknown function specified by sample input/output data pairs is a widespread problem. We already saw in Chapter 2 how multilayer neural networks can implicitly approximate such a mapping. Here we present another approach to this problem by using a neuro-fuzzy system. The discussion that follows is based on the work of D. Nauck [155, 157].

In this approach, called NEFPROX for NEuro Fuzzy function apPROXimator, a neuro-fuzzy systems is seen as a three-layer feedforward network similar to the type described in the preceding section. There are no cycles in the network and no connections exist between layer $n$ and layer $n + j$, with $j > 1$. The first layer represents input variables, the hidden layer represents fuzzy rules, and the third layer represents output variables. The hidden and output units in this network use t-norms and t-conorms as aggregation functions, in a manner similar to what we have seen in the previous sections. Fuzzy sets are encoded as fuzzy connection weights and fuzzy inputs. The whole network is capable of learning and provides a fuzzy inference path. The end result should be interpretable as a system of linguistic rules.

The problem to be solved is that of approximating an unknown continuous function using a fuzzy system given a set of data samples. There is an existence proof that fuzzy systems are capable of universal function approximation [116]. However, actually building such an approximation for a given problem requires the specification of parameters under the form of membership functions and of a rule base. This identification can be done by previous knowledge, trial and error, or by some automatic learning methodology. NEFPROX encodes the problem parameters in the network and uses a supervised learning algorithm derived from neural network theory in order to drive the mapping towards satisfactory solutions. The advantage of the fuzzy approach over a standard neural network is that, while the latter is a black box, the fuzzy system can be interpreted in terms of rules and thus has more descriptive power.

The NEFPROX system is a three-layer network with the following features:

- The input units are labeled $x_1, x_2, \ldots, x_n$. The hidden rule units are called $R_1, R_2, \ldots, R_k$ and the output units are denoted as $y_1, y_2, \ldots, y_m$.
- Each connection is weighted with a fuzzy set and is labeled with a linguistic term.
- All connections coming from the same input unit and having the same label are weighted by the same common weight, which is called a *shared* weight. The same holds for the connections that lead to the same output unit.
- There is no pair of rules with identical antecedents.

According to these definitions, it is possible to interpret a NEFPROX system as a fuzzy system in which each hidden unit stands for a fuzzy if-then rule. Shared weights are needed in order for each linguistic value to have a unique interpretation. If this were not the case, it would be possible for fuzzy weights representing identical linguistic terms to evolve differently during learning, leading to different individual membership functions for its antecedents and conclusions variables, which would in turn prevent proper interpretation of the fuzzy rule base. Figure 6.8 graphically depicts the structure of a NEFPROX system.

Learning in NEFPROX is based on supervised training and employs a conceptual variation of standard backpropagation in ANNs

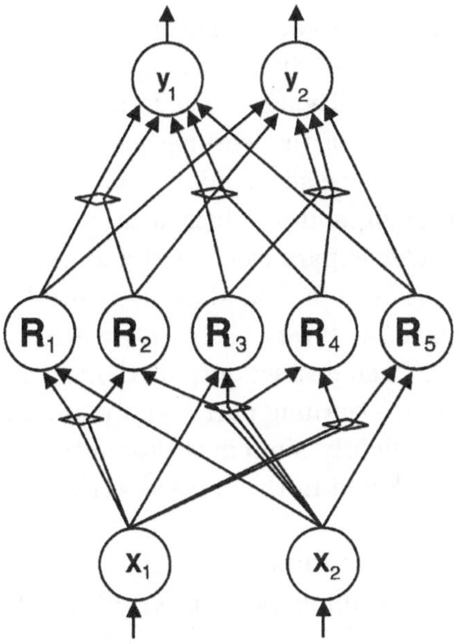


Figure 6.8 Schematic architecture of a NEFPROX system. The connections going through the small rhombuses are linked i.e., they share the same weight (see text). The figure is adapted from Nauck [155].

since we are in a framework where known input/output sets do usually exist. The difference with the standard algorithm is that the method for determining the errors and propagating them backwards to effect local weight modifications is not based on gradient descent. This is due to the fact that the functions involved in the system are not always differentiable, as is the case for some types of triangular norms such as minimum and maximum used here. Central to the NEFPROX approach to learning is simplicity, speed, and interpretability of the results. The system is more suitable for Mamdani-type fuzzy systems with a small number of rules and a small number of meaningful membership functions. Indeed, according to [157], if very precise function approximation is called for, then a neuro-fuzzy approach, which should be characterized by tolerance for imprecision, is probably not well suited anyway and other methodologies should be preferred.

Since fuzzy rules are used in NEFPROX to approximate the unknown function, pre-existing knowledge can be used at the outset by initializing the system with the already known rules, if any. The remaining rules have to be found in the learning process. If nothing is known about the problem, the system starts out without hidden units, which represent rules, and incrementally learns them. This constructive aspect of the algorithm constitutes another difference with respect to the usual backpropagation learning algorithm in a fixed network architecture. Simple triangular membership functions are used for fuzzy sets although other forms would also be permissible. At the beginning of the learning process fuzzy partitions for each input variable are specified. Fuzzy sets for output variables are created during learning and a defuzzyfication procedure is used at the output nodes to compare calculated and observed values. The network structure is as described above (see also Figure 6.8). Given a training set of patterns $\{\mathbf{s}_1, \mathbf{t}_1, \ldots, \mathbf{s}_r, \mathbf{t}_r\}$ where $\mathbf{s} \in \mathbb{R}^n$ is an input pattern and $\mathbf{t} \in \mathbb{R}^m$ the desired output, the learning algorithm has two parts: a structure-learning part and a parameter-learning part. The following is a slightly simplified description of the algorithm, more details can be found in the original work [157].

**Structure Learning Algorithm**

1. Select the next training pattern $(\mathbf{s}, \mathbf{t})$ from the training set.

2. For each input unit $x_i$ find the membership function $\mu_{ji}^{(i)}$ such that
$$\mu_{ji}^{(i)}(s_i) = \max_{j \in \{1,\ldots,p_i\}} \{\mu_j^{(i)}(s_i)\}.$$

3. If there is no rule $R$ with weights $W(x_1, R) = \mu_{ji}^{(1)}, \ldots, W(x_n, R) = \mu_{jn}^{(1)}$ then create the node and connect it to all the output nodes.

4. For each connection from the new rule node to the output nodes find a suitable fuzzy weight $\nu_{ji}^{(i)}$ using the membership functions assigned to the output units $y_i$ such that $\nu_{ji}^{(i)}(t_i) = \max_{j \in \{1,\ldots,q_i\}}\{\nu_j^{(i)}(t_i)\}$ and $\nu_j^{(i)}(t_y) \geq 0.5$. If the fuzzy set is not defined then create a new one $\nu_{\text{new}}^{(i)}(t_i)$ for the output variable $y_i$ and set $W(R, y_i) = \nu_{\text{new}}^{(i)}$.

5. If there are no more training patterns then stop rule creation; otherwise go to 1.

6. Evaluate the rule base and change the rule conclusions if appropriate.

The supervised learning part that adapts the fuzzy sets associated to the connection weights works according to the following schema:

**Parameter Learning Algorithm**

1. Select the next training pattern $(\mathbf{s}, \mathbf{t})$ from the training set and present it at the input layer.

2. Propagate the pattern forward through the hidden layer and let the output units determine the output vector $\mathbf{o}$.

3. For each output unit $y_i$ determine the error $\delta_{yi} = t_i - o_{yi}$.

4. For each rule unit $R$ with output $o_R > 0$ do:

   - Update the parameters of the fuzzy sets $W(R, y_i)$ using a learning rate parameter $\sigma > 0$.
   - Determine the change $\delta_R = o_R(1 - o_R) \cdot \sum_{y \in \text{output layer}}(2W(R, y)(t_i) - 1) \cdot |\delta_y|$.
   - Update the parameters of the fuzzy sets $W(x, R)$ using $\delta_R$ and $\sigma$ to calculate the variations.

5. If a pass through the training set has been completed and the convergence criterion is met then stop; otherwise go to step 1.

The learning procedure for the fuzzy sets is based on simple heuristics that result in shifting the membership functions and in making their support larger or smaller. It is possible and easy to impose constraints on the learning procedures such as that fuzzy sets must not pass each other or that they must intersect at some point and so on. As usual in supervised learning algorithms, one or more validation sets of data are used and training goes on until the error on the validation set starts to increase in order to avoid overfitting and to promote generalization (see also Chapter 2, Section 2.5).

NEFPROX has been tested on a well-know difficult benchmark problem: the Mackey-Glass system. The Mackey-Glass delay-differential equation was originally proposed as a model of white blood cell production:

$$\frac{dx}{dt} = \frac{0.2x(t-\tau)}{1 + x^{10}(t-\tau)} - 0.1x(t), \tag{6.14}$$

where $\tau$ is a parameter. Using a value of $\tau = 17$ the resulting series is chaotic. Training data can be obtained by numerical integration of the equation. A thousand values were calculated of which the first half were used for training and the rest for validation. The NEFPROX system used to approximate the time series has four input and one output variable and each variable was initially partitioned by seven equally distributed fuzzy sets with neighboring membership functions intersecting at degree 0.5. After learning, 129 fuzzy rules were created. The resulting system approximates the function quite well in the given range. The results are only slightly worse than those that have been obtained on the same problem with another neuro-fuzzy system called ANFIS [103] (see next section) but the learning time is much shorter.

Two related neuro-fuzzy approaches are NEFCON and NEFCLASS which are used, respectively, for control applications and for classification problems [155]. NEFCON is similar to NEFPROX but has only one output variable and the network is trained by reinforcement learning using a rule-based fuzzy error measure as a reinforcement signal. NEFCLASS sees pattern classification as a special case of function approximation and uses supervised learning in a manner similar to NEFPROX to learn classification rules.

### 6.2.5 A Second Example: The ANFIS System

ANFIS stands for Adaptive Network-based Fuzzy Inference System and is a neuro-fuzzy system that can identify parameters by using supervised learning methods [103]. ANFIS can be thought of as a network representation of Sugeno-type fuzzy systems with learning capabilities. ANFIS is similar in spirit to NEFPROX but, with respect to the latter, learning takes place in a fixed structure network and it requires differentiable functions. The ANFIS heterogeneous network architecture is constituted by a number of layers of nodes which have the same function for a given layer but are different from one layer to the next. For example, consider the fuzzy inference system with two inputs $x$ and $y$ and a single output $z$ [103]. For a first-order Sugeno model, a rule set using a linear combination of the inputs can be expressed as:

$$\begin{array}{l} \text{IF } x \text{ is } A_1 \text{ AND } y \text{ is } B_2 \text{ THEN } f_1 = p_1 x + q_1 y + r_1 \\ \text{IF } x \text{ is } A_2 \text{ AND } y \text{ is } B_2 \text{ THEN } f_2 = p_2 x + q_2 y + r_2 \end{array} \tag{6.15}$$

The reasoning mechanism for this model is:

$$f = \frac{w_1 f_1 + w_2 f_2}{w_1 + w_2} = \bar{w}_1 + \bar{w}_2. \tag{6.16}$$

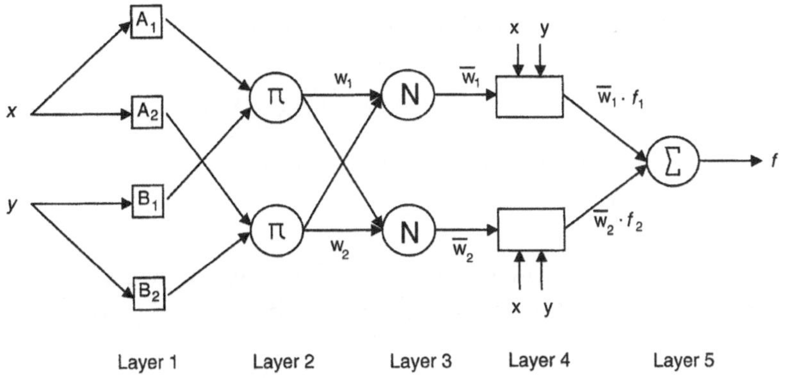


Figure 6.9 ANFIS architecture corresponding to a two-input first-order Sugeno fuzzy model with two rules (see text). The figure is adapted from the work of Jang and Sun [103].

The ANFIS network architecture corresponding to this Sugeno model is shown in Figure 6.9. The layers in the net are constituted by nodes having the same function for a given layer. The functionalities of the layers are as follows:

Layer 1: Denoting by $O_{l,i}$ the output of node $i$ in layer $l$, each node in layer 1 is an adaptive unit with output given by:

$$O_{1,i} = \mu_{A_i}(x), \quad i = 1, 2 \tag{6.17}$$
$$O_{1,i} = \mu_{B_{i-2}}(x), \quad i = 3, 4$$

where $x$ and $y$ are input values to the node and $A_i$ or $B_{i-2}$ are fuzzy sets associated with the node. In other words, each node in this layer generates the membership grades of the premise part. The membership functions for $A_i$ and $B_i$ can be any appropriate parameterized membership function such as triangular, trapezoidal, Gaussian or bell-shaped.

Layer 2: Each node in this layer is labeled Π and computes the firing strength of each rule as the product of the incoming inputs or any other t-norm operator:

$$O_{2,i} = w_i = \mu_{A_i}(x) \bigtriangleup \mu_{B_i}(y), \quad i = 1, 2. \tag{6.18}$$

Layer 3: Each node in this layer is labeled N and it calculates the ratio of the i-th rule's firing strength to the sum of all rules' firing strengths:

$$O_{3,i} = \bar{w}_i = \frac{w_i}{w_1 + w_2}, \quad i = 1, 2. \tag{6.19}$$

Layer 4: Each node in this layer has the following function:

$$O_{4,i} = \bar{w}_i f_i = \bar{w}_i(p_i x + q_i y + r_i), \tag{6.20}$$

where $\bar{w}_i$ is the output of layer 3 and $\{p_i, q_i, r_i\}$ is the parameter set (see Equation 6.15).

Layer 5: There is a single node $\Sigma$ in this layer. It aggregates the overall output as the summation of all the incoming signals:

$$O_{5,1} = \sum_i \bar{w}_i f_i = \frac{\sum_i w_i f_i}{\sum_i w_i} \tag{6.21}$$

This completes the construction of the network which is seen to have the same functionality as the equivalent Sugeno model.

### Learning in ANFIS

The ANFIS learning algorithm is a hybrid supervised method based on gradient descent and least-squares methods. In the forward phase, signals travel forward up to layer 4 and the relevant parameters are fitted by least-squares. In the backward phase the error signals travel backward and the premise parameters are updated as in backpropagation. More details of the algorithm can be found in [103]. It is worth noting that the ANFIS network with its learning capabilities can be built by using the fuzzy toolbox available in the MATLAB package.

### Function Modeling and Time Series Prediction

ANFIS can be applied to non-linear function modeling and time series prediction. ANFIS gives excellent results on the prediction of the time series generated by the numerical integration of the Mackey-Glass delay-differential equation prediction of the time series generated by the numerical integration of this equation, better than most other approaches for function approximation such as those based on neural networks of various types [103] and on standard function fitting methods. The ANFIS system shows excellent non-linear fitting and generalization capabilities on this example. As well, the number of parameters and the training time is comparable or less than what is required by ANN methods, with the exception of the neuro-fuzzy system NEFPROX, which learns faster and has slightly fewer parameters, as we saw above.

### ANFIS for Neuro-Fuzzy Control

Fuzzy control has been introduced in Chapter 3, Section 5.2.4. The time evolution of a dynamical system can be described by the following differential equation:

$$\frac{d\mathbf{x}}{dt} = F(\mathbf{x}, \mathbf{u}),$$

where $\mathbf{x}$ represents the state of the system and $\mathbf{u}$ is a vector of controllable parameters. The control action is formally given by a function $g$ that maps the system state into appropriate parameters for a given control problem:

$$\mathbf{u}(t) = g(\mathbf{x}(t)).$$

We have seen in Chapter 3, Section 5.2.4 that the problem of finding optimal control policies for non-linear systems is mathematically very difficult, while fuzzy approaches have proved effective in many cases. Since a wide class of fuzzy controllers can be transformed into equivalent adaptive networks, ANFIS can be used for building intelligent controllers that is, controllers that can reason with simple fuzzy inference and that are able to learn from experience in the ANN style.

## 6.3 "Co-operative" Neuro-fuzzy Systems

Another level of integration between artificial neural networks and fuzzy systems tries to take advantage of the array of adaptation and learning algorithms devised for the former to tune or create all or some aspects of the latter, and *vice versa*.

One important thing to note is that such approaches refrain from casting the fuzzy system into a network structure, or fuzzifying the elements of the neural network, unlike other approaches discussed in Section 6.2.

One could also note that, under this perspective, radial-basis function networks, a type of neural network with bell-shaped activation functions instead of sigmoid, might be interpreted as neuro-fuzzy networks in their own way, simply by considering their activation functions as membership functions.

### 6.3.1 Adaptive Fuzzy Associative Memories

A possible interpretation of a fuzzy rule, proposed by Kosko [115], views it as an association between antecedent and consequent variables. Kosko calls a fuzzy rule base complying with that semantic interpretation a *fuzzy associative memory* (FAM).

#### Associative Memories

An associative memory consists of memory components of the form $(k, i)$, where $k$ is a key and $i$ the information associated with it. Retrieval of a memory component depends only on its key and not on its place in the memory. Recall is done by presenting a key $k^*$, which is simultaneously compared to the keys of all memory components. The information part $i^*$ is found (or reported missing) within one memory cycle.

Associative memories can be implemented as neural networks, and in that case one speaks of neural associative memories: if a key pattern is presented to a neural associative memory, the activations of the output units represent the corresponding information pattern.

#### Fuzzy Associative Memories

When a variable $x$ takes up values in a finite discrete domain $X = \{x_1, \ldots, x_m\}$, a fuzzy set $A$ with membership function $\mu_A\colon X \to [0,1]$ can be viewed as a point $\mathbf{v}_A$ in the $m$-dimensional hypercube, identified by the co-ordinates

$$\mathbf{v}_A = (\mu_A(x_1), \ldots, \mu_A(x_m)).$$

Accordingly, a fuzzy rule $R$ of the form

$$\text{IF } x \text{ is } A \text{ THEN } y \text{ is } B$$

can be viewed as a function mapping $\mathbf{v}_A$ (a point in $[0,1]^m$) to $\mathbf{v}_B$ (a point in the hypercube, say $[0,1]^s$ defined by the domain $Y$ of $y$).

A fuzzy associative memory is a two-layer network, with one input unit for each discrete value $x_i$ in every domain $X$ of input variables and one output unit for each discrete value $y_j$ in the output variable domain $Y$. Activation for all units can range in $[0,1]$ and is to be interpreted as the degree of membership of the relevant discrete value in the relevant linguistic value. The weights between input unit-output unit pairs can range in $[0,1]$ and the activation function for output unit $u_j$ is

$$u_j = \mathop{\nabla}_{i=1,\ldots,m} v_i \vartriangle w_{ij}. \tag{6.22}$$

A FAM is determined by its connection weight matrix $\mathbf{W} = (w_ij)$, with $i = 1, \ldots, m$ and $j = 1, \ldots, s$. Such a FAM stores just one rule. Matrix $\mathbf{W}$ is called *fuzzy Hebb matrix*.

Given an input fuzzy set $A$ in the form of a vector $\mathbf{v}_A$ and the corresponding output fuzzy set $B$ in the form of a vector $\mathbf{u}_B$, the fuzzy Hebb matrix storing their association is given by the *correlation minimum encoding* [115]

$$\mathbf{W} = \mathbf{v} \circ \mathbf{u}, \quad w_{ij} = v_i \vartriangle u_j. \tag{6.23}$$

The associative recall is given by

$$\mathbf{u} = \mathbf{v} \circ \mathbf{W}, \quad u_j = \mathop{\nabla}_{i=1,\ldots,m} v_i \vartriangle w_ij. \tag{6.24}$$

The recall is always correct if $h(\mu_A) \geq h(\mu_B)$, where $h(\cdot)$ is the height of a membership function, i.e., the maximum degree of membership. If we restrict our attention to normal fuzzy sets, then the recall will always be correct.

Summarizing, the concept of a FAM should be nothing really new to the reader. Once the notational details are clear, one can recognize that a FAM is simply a matrix-vector representation of a fuzzy relation or a fuzzy rule, and one which resembles very closely two-layer neural networks.

### FAM Systems

Because combination of multiple fuzzy Hebb matrices into a single matrix is not recommended lest a severe loss of information is incurred, each rule of a fuzzy rule base should be represented by a distinct FAM. The overall output of the system is then given by the component-wise maximum of all FAM outputs. Such a fuzzy system, shown in Figure 6.10, is called a *FAM system*.

The FAM system is completed by a fuzzification and a defuzzification component, and by weights associated with each FAM.

One strength of a FAM's striking resemblance with a two-layer artificial neural network is that we can borrow some learning techniques and make FAMs adaptive.

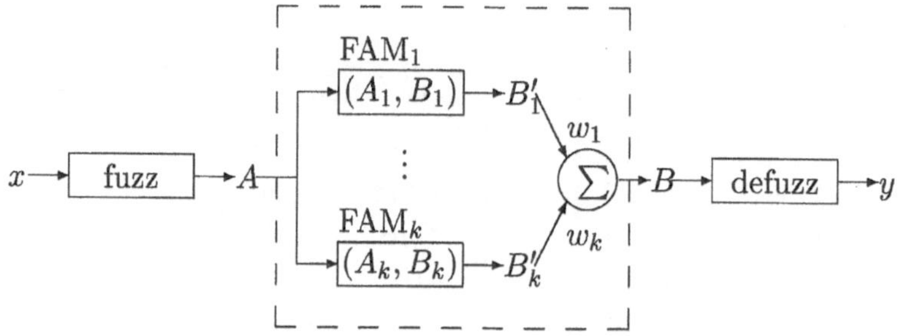


Figure 6.10 General scheme of a FAM system.

## Learning in Adaptive FAMs

Kosko suggests two approaches to learning for adaptive FAMs. The first possibility is to learn the weights associated with FAMs' outputs. The second and more interesting is to create FAMs completely by learning.

The learning procedure proposed by Kosko is *differential competitive learning*, a form of adaptive vector quantization (see Chapter 2, Section 2.6.2).

Given a data set of examples of inputs and correct output values, with $n$ input variables and one output variable, the idea is to build a two-layer network with as many input units as variables (i.e., $n + 1$) and as many output units as the number of possible rules one could build from the given variable and a predefined partition of their domains (that is, the user must determine the linguistic values in advance). To begin with, all the input units are connected to all the output units and the output units are completely connected by inhibitory links. The examples in the data set form clusters in the product space of all variables; the learning process is supposed to develop prototypes of these clusters; each cluster is interpreted as an instance of a fuzzy rule and their best matching prototypes are selected by the learning process. Therefore, the learning procedure selects the FAMs to be included in the system and assigns them a weight; if further training data are collected during the use of the FAM system, the learning process can resume and continue concurrently with the operation of the system, by updating the rule weights or by deleting or adding rules.

### 6.3.2 Self-Organizing Feature Maps

An approach similar to Kosko's differential competitive learning is proposed by Pedrycz and Card [171]. They use a self-organizing feature map (cf. Chapter 2, Section 2.6.3) with a planar competition layer to cluster training data, and they provide means to interpret the result of learning as linguistic rules.

The self-organizing map has an input layer with $n$ units, where $n$ is the number of variables in a dataset record. The output layer of the map is a $n_1 \times n_2$ lattice of units. Inputs and connection weights are in $[0, 1]$. It is convenient to specify the connection weights between input and output units as a three-dimensional matrix $\mathbf{W} = (w_{i_1,i_2,i})$, where $i_1 = 1, \ldots, n_1$, $i_2 = 1, \ldots, n_2$, and $i = 1, \ldots, n$.

The result of learning a set of sample records (or vectors) $\mathbf{x}_k = (x_{k1}, \ldots x_{kn})$, $k = 1, \ldots, m$, shows whether two input records are similar, i.e., belong to the same class. However, if $n$ is sufficiently large, the structure of the problem is not usually detected in the two-dimensional map. Rather, Pedrycz and Card provide a procedure for interpreting the result using linguistic variables.

After learning, each variable $x_i$ can be described by a matrix $\mathbf{W}_i$, which contains the weights of the connections between the relevant input unit $u_i$ and all the output units. This constitutes the map for a single variable, or feature. The procedure consists in specifying a number of fuzzy sets $A^{(i)}_{j_i}$ for each variable $x_i$, with membership function $\mu_{A^{(i)}_{j_i}}$. These membership functions are applied to matrix $\mathbf{W}_i$ to obtain an equal number of transformed matrices $\mu_{A^{(i)}_{j_i}}(\mathbf{W}_i)$. The transformed matrices have higher values in those areas of the map that are compatible with the linguistic concept represented by $A^{(i)}_{j_i}$.

Each combination of linguistic terms is a possible linguistic description of a cluster of records from the data set. To check a linguistic description for validity, the transformed matrices are intersected, yielding a matrix $\mathbf{D} = (d_{i_1,i_2})$, which can be interpreted as a fuzzy relation among the variables:

$$\mathbf{D} = \bigwedge_{i=1}^{n} \mu_{A^{(i)}_{j_i}}(\mathbf{W}_i), \quad d_{i_1,i_2} = \min_{i=1,\ldots,n} \{\mu_{A^{(i)}_{j_i}}(w_{i_1,i_2,i})\}. \tag{6.25}$$

Each linguistic description is a valid description of a cluster if the relevant fuzzy relation $\mathbf{D}$ has a non-empty $\alpha$-cut $\mathbf{D}_\alpha$. If the variables are separated into input and output variables, according to the particular problem at hand, then each valid linguistic description readily translates into an IF-THEN rule.

Compared to Kosko's FAMs, Pedrycz and Card's approach is more computationally expensive, because all combinations of linguistic terms must be examined in order to produce the desired fuzzy rule base. Furthermore, the determination of a sufficiently high threshold $\alpha$ for assessing description validity and of the right number of neurons in the output layer is a problem that has to be individually solved for every learning problem. However, the advantages are that the rules are not weighted and that the user-defined fuzzy sets have a guiding influence on the learning process, leading to more interpretable results.

### 6.3.3 Learning Fuzzy Sets for Sugeno-Type Fuzzy Systems

A method for learning the fuzzy sets in a Sugeno-type fuzzy rule-based system using supervised learning is the one proposed by Nomura and colleagues [169].

First of all, the assumption is made that linguistic values referred to by rule antecedents are defined by parameterized triangular fuzzy numbers, that is, membership functions of the form

$$\mu(x) = \begin{cases} 1 - 2\frac{|x-C|}{b} & C - \frac{b}{2} \leq x \leq C + \frac{b}{2} \\ 0 & \text{otherwise,} \end{cases} \tag{6.26}$$

where $C$ is the center and $b$ is the base, or width, of the triangle. The consequent of a rule just consists of a crisp value $w_0$ (this is a degenerate case of the Sugeno model). The product is used as the t-norm.

The learning algorithm is based on gradient descent using the half squared error as the error measure:

$$\text{hse} = \frac{1}{2}\sum_{k=1}^{m}(y_k - y_k^*)^2, \tag{6.27}$$

where $y_k$ is the value computed by the fuzzy system and $y_k^*$ is the actual value in the training data set. Since the type of Sugeno model adopted applies only differentiable operations, both to determine the degree of truth of the antecedents (product) and to aggregate the outputs of all the rules (weighted average), the calculation of the changes for the parameters $C$ and $b$ of each membership function and $w_0$ of each rule is equivalent to the generalized delta rule (cf. Chapter 2, Section 2.5) for multilayer neural networks.

The only caution one must have is that the triangular membership functions are not differentiable in three points. However, it is not too difficult to devise satisfactory heuristics that overcome this potential problem.

One disadvantage of this approach is that the semantics of the linguistic values depend on the rules they appear in. Whereas in the initial (hand-crafted) rule base identical linguistic terms are described by distinct yet identical membership functions, the learning procedure changes this state of affairs, by modifying the membership functions of each term independently of the others. Such effect is undesirable, for it obfuscates the interpretation of the resulting rule base. A way to overcome this difficulty, proposed by Bersini, Nordvik, and Bonarini [27] would be to make identical linguistic terms share the same membership function.

### 6.3.4 Fuzzy ART and Fuzzy ARTMAP

A Fuzzy ART (for "Adaptive Resonance Theory") neural network [41] is a self-organizing neural network capable of clustering collections of arbitrarily complex analog input patterns via unsupervised learning.

#### Fuzzy ART Neural Networks

The Fuzzy ART neural network architecture, illustrated in Figure 6.11, consists of two subsystems, the *attentional* subsystem and the *orienting* subsystem. The attentional subsystem consists of two layers $L_1$ and $L_2$. $L_1$ is called the *input* layer because input patterns are applied to it; $L_2$ is called the *category* or *class representation* layer because it is the layer where category representations, i.e., the clusters to which input patterns belong, are formed. The orienting subsystem consists of a single node (the *reset* node), which accepts inputs from the nodes in $L_1$ and $L_2$ and the input pattern directly; its output affects the nodes in $L_2$.

Patterns are assumed to be $n$-dimensional vectors in $[0,1]^n$; the input to the Fuzzy ART network is formed by putting the pattern and its complement in a $2n$-dimensional vector $\mathbf{x}$.

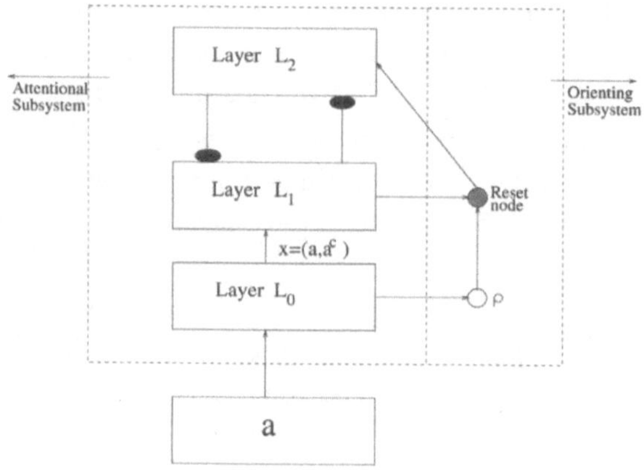


Figure 6.11 Network architecture of a Fuzzy ART.

Each category $j$ in $L_2$ corresponds to a vector $\mathbf{w}_j = (w_{j1}, \ldots, w_{j,2n})$ of weights. All these weights are initially set to one: such a category is said to be *uncommitted*. After a category has been chosen to represent an input pattern, it is referred to as a *committed* category or node.

Training a Fuzzy ART means tuning its weights so as to cluster the input patterns $\mathbf{x}_1, \ldots, \mathbf{x}_P$ into different categories, according to their similarity. The dataset of patterns is repeatedly presented to the Fuzzy ART in the given order, as many times as necessary. Training is considered accomplished when the weights do not change during a complete dataset presentation. This training scenario is called *off-line* training.

Off-line training proceeds as follows: each time an input pattern $\mathbf{x}_i$ is presented, the input $T_j(\mathbf{x}_i)$ to each category node $j$ is calculated as

$$T_j(\mathbf{x}_i) = \begin{cases} \frac{n}{\alpha+2n}, & \text{if } j \text{ is uncommitted;} \\ \frac{|\mathbf{x}_i \wedge \mathbf{w}_j|}{\alpha+|\mathbf{w}_j|}, & \text{if } j \text{ is committed.} \end{cases} \qquad (6.28)$$

Now, let's call $j^*$ the node in $L_2$ receiving the maximum input from $L_1$,

$$j^* = \arg\max_{j \in L_2} T_j(\mathbf{x}_i).$$

Two cases require special actions:

1. if node $j^*$ is uncommitted, a new uncommitted node in $L_2$ is introduced, and its weights are all initialized to one;

2. if node $j^*$ is committed but

$$\frac{|\mathbf{x}_i \wedge \mathbf{w}_j|}{|\mathbf{x}_i|} < \rho,$$

(i.e., it does not satisfy the vigilance criterion), it is disqualified by setting $T_{j^*}(\mathbf{x}_i) = -1$ and the next maximum-input node in $L_2$ is considered, until either case 1 is verified or $j^*$ satisfies the vigilance criterion.

At this point, the weights associated with node $j^*$ are modified according to the equation

$$\mathbf{w}_{j^*} \leftarrow \mathbf{w}_{j^*} \wedge \mathbf{x}_i.$$

Quantity $\rho \in [0, 1]$ is the *vigilance* parameter, affecting the resolution of clustering: small values of $\rho$ result in coarse clustering, whereas larger values result in finer clustering of input patterns. Parameter $\alpha \in (0, +\infty)$ is called the *choice* parameter.

The $\wedge$ operation in the above equations is the componentwise minimum, and could be replaced in principle by any t-norm.

### Fuzzy ARTMAP

A Fuzzy ART module generates the categories needed to classify the input by unsupervised learning, according to a similarity criterion. A composition of Fuzzy ART modules makes up a Fuzzy ARTMAP [40], a neuro-fuzzy architecture capable of learning the relationship between data and user-defined categories (supervised learning).

A Fuzzy ARTMAP, illustrated in Figure 6.12, consists of two Fuzzy ART modules, $\text{ART}_a$ and $\text{ART}_b$, plus a MAP module. $\text{ART}_a$ receives in input the patterns to be classified; $\text{ART}_b$ receives in input the $m$ user-defined classes to which the patterns belong, and generates an internal category layer corresponding to it. The MAP module connects the two Fuzzy ART modules and tries to minimize the difference between the classes generated by $\text{ART}_a$ from data and the user-defined classes generated by $\text{ART}_b$. In other words, the MAP module builds a mapping from the $n$ inputs to the $m$ classes, possibly by acting on the vigilance parameter $\rho_a$ of $\text{ART}_a$.

Among the main features of Fuzzy ART and Fuzzy ARTMAP are the minimization of the predictive error, the maximization of code compression and generalization, the dynamic introduction of new categories when needed, the possibility of distinguishing exceptions from noise, and fast convergence. All these features make this type of neuro-fuzzy architecture very popular in a variety of applications.

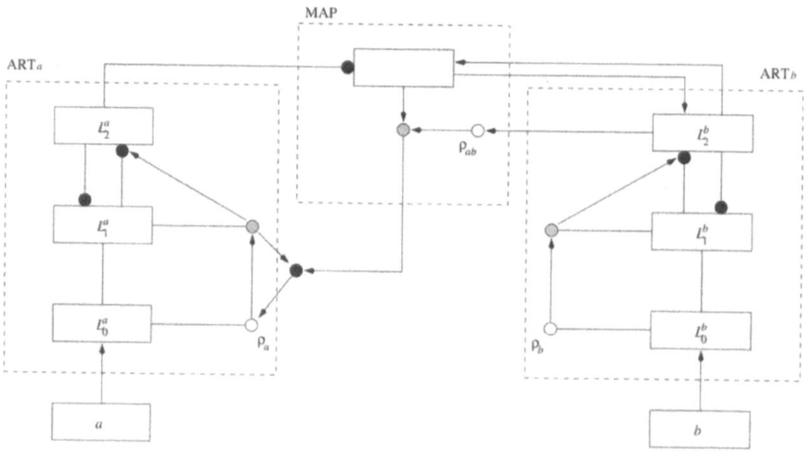


Figure 6.12 Network architecture of a Fuzzy ARTMAP.

## 6.4 Applications of Neuro-fuzzy Systems

NEURO-FUZZY SYSTEMS are passing from their infancy, where research on foundations and methodologies predominates over applications, to maturity, where principles and methodologies become technology deployed on the field. This passage has been fostered by several methodological developments that have been described in the previous sections. In this last section, which roughly follows Chapter 14 of [223], we give an overview of neuro-fuzzy systems applications that have gained their acceptance on the field, without any ambition of being comprehensive. In fact, new valuable applications appear every year, and any attempt at making a survey would soon become obsolete.

### 6.4.1 Engineering

Neuro-fuzzy approaches have been used in a variety of engineering applications, including consumer electronics, control, diagnostics, manufacture, biotechnology, power generation, chemical processes, power electronics, communications, and software resource management. It is now rather well established that neuro-fuzzy systems can adequately adapt to changing environmental conditions.

Ishibuchi and colleagues [102] applied an ANFIS-like fuzzy neural network (Sugeno-type fuzzy rules with gradient descent learning) to rice tasting, which involves the development of a six-variable fuzzy relation.

Of particular interest to our synergetic vision of soft computing is the work reported by Pao [162], where neural networks, fuzzy logic, and evolutionary algorithms are combined to support the task of process monitoring and optimization in electric power utilites, including heat rate improvement and NO emission minimization.

### 6.4.2 Diagnostics in Complex Systems

Neuro-fuzzy systems have been applied to several problems arising in the aerospace industry, like control surface failure detection for a high-performance aircraft [184]. The detection model is developed using a linear dynamic model of an F-18 aircraft. The detection scheme makes use of a residual tracking error between the actual system and the model output to detect and identify a particular fault. Two parallel models detect the existence of a surface failure, whereas the isolation and magnitude of any one of the possible failure modes is estimated by a neuro-fuzzy decision algorithm. Simulation results demonstrate that detection can be achieved without false alarms even in the presence of actuator/sensor dynamics and noise.

Typical examples of complex systems whose modeling and monitoring is of critical importance are nuclear reactors. Several applications of neuro-fuzzy techniques have been described in the literature [224].

### 6.4.3 Control

Neuro-fuzzy systems have found broad application in control, probably more so than in any other field. Controlled plants are as diverse as industrial sewing machines [213] and fusion reactors [250], home electric appliances [236] like refrigerators, air-conditioning systems, and welding machines, and consumer electronic devices such as hand-held video cameras.

### 6.4.4 Robotics

In the field or robotics, neuro-fuzzy systems have been employed for supervisory control, planning, grasping, and guidance. Grasping

[61] has to do with the control of robotic arms with three or more fingers; the main issue is finding an optimal coordination of the forces applied to the object, in order to hold it firmly without squeezing or damaging it. On-line learning ensures that grasp parameters are continuously adjusted to current conditions.

Approaches based on Fuzzy ART have been used for autonomous robot guidance and navigation: for instance, Bonarini and Ludovico [30] report on a system that is able to learn a topological map derived from different robot sensors, and use it to navigate in unstructured environments. Their approach is based on the definition and identification of grounded concepts by integrated instances of Fuzzy ART.

### 6.4.5 Image Processing

Image processing and pattern recognition are a field in which neural networks play a prominent role, especially because it naturally lends itself to massively parallel and connectionist processing of the type supported by ANNs. However, in recent years, neuro-fuzzy systems have been designed and investigated to improve on ANN performance in image processing tasks both at low level, such as in image quality enhancement and image manipulation, and high level, such as in edge detection, pattern recognition, medical and environmental imaging.

### 6.4.6 Finance

ANNs have been used for years by the financial community in investment and trading because of their ability to identify patterns of behavior that are not readily available and complex relations among variables that are hidden by chaos and noise. Much of this work has never been published for obvious reasons, although generic and casual accounts of it have been given.

Since the mid 1990s, neuro-fuzzy techniques have been incorporated into some financial applications. For instance, a neuro-fuzzy decision support system is described in [96], which tries to determine whether a given stock is underpriced or overpriced by learning patterns associated with either condition.

Another neuro-fuzzy system [39] is capable of evaluating portfolios by using currency exchange rate fluctuations and expert knowledge given in the form of fuzzy IF-THEN rules; the merit of the neuro-fuzzy system is its ability to automatically fine tune the expert knowledge with a backpropagation-like algorithm.

Chapter 7

# Fuzzy Evolutionary Algorithms

## 7.1 Introduction

SYNERGY between evolutionary algorithms and fuzzy logic can occur in three complementary forms [220].

The most obvious form exploits the optimum searching ability of evolutionary algorithms to synthesize and optimize fuzzy systems. This was the object of Chapter 5.

As for the second possibility, evolutionary algorithms are relatively easy to implement and in general their performance tends to be rather satisfactory in comparison with the small amount of knowledge about the problem they need in order to work; however, typically they would require some sort of human supervision during their use as practical tools. The idea is then to use a fuzzy knowledge base (or *fuzzy government* [8]) to detect the emergence of a solution, dynamically tune algorithm parameters, as in [122] or [26], and monitor the evolutionary process to avoid undesired behaviors such as premature convergence.

A third option is to embed fuzziness into the evolutionary algorithm itself. For instance, some precision in the calculation of fitness could be sacrificed to save computational resources by defining a fuzzy fitness, or even fuzzy alleles for genes, as in [200], thus fuzzifying also genetic operators.

This chapter concentrates on these two latter forms, whose distinctive feature could be stated as the use of fuzzy tools or fuzzy-logic-based techniques to model different components or control the evolutionary process, with the goal of improving performance [91].

## 7.2 Fuzzy Control of Evolution

THE SETTING of evolutionary algorithm parameters ia a key factor in the determination of the exploitation versus exploration trade-off. It has long been acknowledged that parameter setting has a significant impact on evolutionary algorithm performance [80]. The effects of poor settings can range from a slow-down of the search process to premature convergence. However, finding robust control parameters valid for any problem is difficult and perhaps even impossible. As a matter of fact, their interaction with the evolutionary machinery is complex and optimal parameters are problem-dependent [13]. Furthermore, different settings of control parameters may be optimal in different moments of the same evolutionary process.

For this reason, an evolutionary algorithm typically requires some sort of human supervision during its use to solve real world problems. A human supervisor can observe the process and tweak parameters as evolution proceeds until a satisfactory behavior is obtained. Finally, there is the problem of deciding when to stop the search and be content with the solution produced by the evolutionary process. This is in general the result of a trade-off between solution quality and time constraints and another area of discretionary human intervention.

The alternative to human supervision is an *adaptive* evolutionary algorithm, capable of adjusting its own control parameters during evolution to attain the best performance. Starting from the observation that the evolutionary process is a complex system with non-linear behavior, it seems sensible to take advantage of a powerful technique such as fuzzy control (cf. chapter 5, section 5.2.4) to manage it.

### 7.2.1 Fuzzy Government

Following [8], the suggestive name of *Fuzzy Government* can be given to the collection of fuzzy rules and routines in charge of controlling the evolution of a population, detecting the emergence of a solution, tuning algorithm parameters "on flight" (cf. [122]), and avoiding undesirable behaviors such as premature convergence.

A fuzzy controller implements an expert operator's approximate reasoning process in the selection of a control action. The fuzzy government dynamically adjusts selected control parameters or genetic operators during operation of an evolutionary algorithm in order to offer the most appropriate exploration and exploitation behavior.

The name is motivated by the similarity with the ideal role of the government of a nation: at least in principle, a government should employ its limited powers to enforce the law and bring about favorable conditions for the prosperity and well-being of its subjects. In the same fashion, the fuzzy government operates on the population of individuals of an evolutionary algorithm so that the most favorable conditions for the improvement of their fitness are always maintained.

The attractiveness of using control techniques based on Fuzzy Set Theory (see Chapter 3) is that they allow the human expert (the algorithm developer) to encapsulate their somehow imprecise knowledge about the system (the evolutionary process) in an easy and straightforward way.

### 7.2.2 Architecture of an Adaptive Evolutionary Algorithm

The main idea of an adaptive evolutionary algorithm, sketched in Figure 7.1, is to use a fuzzy rule-based controller, the fuzzy government, whose inputs are statistics of the evolutionary algorithm and whose outputs are the control parameters of the same algorithm.

Statistics are gathered from the evolutionary algorithm at a given sampling rate, for instance once in $r$ generations, but anyway not more than once per generation, and they are sent to the fuzzy government. The fuzzy government performs one inference, producing new values for all control parameters. These are updated with the new values and control is given back to the evolutionary algorithm.

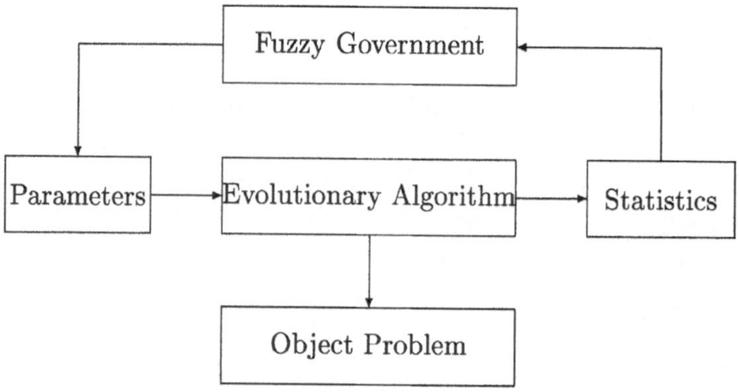


Figure 7.1 The abstract architecture of an adaptive evolutionary algorithm.

The overhead introduced by the fuzzy government is in general negligible, and can be made as small as one wishes just by decreasing the sampling rate $1/r$. In exchange, experience has demonstrated that improvements in performance can be dramatic.

**Evolutionary Algorithm Statistics**

Statistics of an evolutionary algorithm can be divided into two classes [38]: genotypic statistics, which summarize aspects related to the genotypes of individuals in a population, regardless of their meaning when decoded, and phenotypic statistics, which concern properties of individual performance for the problem at hand, or fitness.

Diversity measures constitute a general kind of genotypic statistics. They rely on the definition of a (fuzzy) similarity measure (cf. Chapter 3, Section 3.3.3)

$$\mu_{\text{similar}}(\gamma, \kappa),$$

where $\gamma$ and $\kappa$ are two genotypes.

Other kinds of genotypic statistics may be defined more or less in an *ad hoc* fashion, depending on particular features of an evolutionary algorithm. For instance, in an algorithm using a repair function, one could define a statistics on the average number of times that function is invoked per generation.

Examples of phenotypic statistics are: phenotypic diversity measures (such as fitness range, best/average fitness ratio, or fitness variance), maximum, average, and minimum fitness.

As in the case of genotypic statistics, particular statistics may be defined relevant to some specific features of the object problem, of the evolutionary algorithm implementation, or both.

Other statistics, which are less easy to classify, may finally be meaningful in some settings. For instance, success rate could be defined, following [185] (P. 122) as the ratio of mutations that end up in a fitness improvement for the individual undergoing them over the total number of mutations. This allows one to take into account the famous $\frac{1}{5}$-success rule of Evolution Strategies [185].

In [122] and [247], it is suggested that current values of some control parameters may also be used as inputs to the fuzzy government.

#### Control Parameters

Outputs of the fuzzy government may be directly control parameter values or changes in these parameters [122]. Control parameters may be the usual parameters of an evolutionary algorithm, such as population size, mutation and crossover rate, selection pressure (if applicable), or other parameters specific to a particular application.

Among control parameters one could also include indicators for the user. These "improper" parameters tell the user of the evolutionary algorithm something about the state of optimization. For instance, one such indicator could be the degree to which a satisfactory solution has been obtained.

#### The Fuzzy-Rule-Base

Each input and output variable is to be associated with a set of linguistic values, defined by membership functions. The fuzzy rules in the fuzzy government are defined in terms of such linguistic values.

Different approaches to the construction of a suitable fuzzy-rule-base are found in the literature. One alternative is to study the behavior of the evolutionary algorithm and gain some expertise about its dynamics. This expert knowledge, coupled with general knowledge about evolutionary algorithms, may then be sufficient to craft suitable rules. Alternative approaches involve automatic learning of the fuzzy rules, membership functions, or both.

### 7.2.3 Using Expert Knowledge

In [247, 248], a fuzzy government is used to solve two problems evolutionary algorithms may incur: very slow convergence speed and premature convergence. These problems may be due to at least three factors:

| $p_c$ | Population Size | | |
|---|---|---|---|
| Generation | small | medium | large |
| short | medium | small | small |
| medium | large | large | medium |
| long | very large | very large | large |

Table 7.1 Fuzzy rule-bases for the control of crossover rate and mutation rate respectively, suggested by Xu and colleagues [247, 248].

| $p_m$ | Population Size | | |
|---|---|---|---|
| Generation | small | medium | large |
| short | large | medium | small |
| medium | medium | small | very small |
| long | small | very small | very small |

1. parameters are not well chosen initially for a given task;
2. parameters do not change through a run, even though evolutionary conditions may vary;
3. interaction between distinct parameters is complex and difficult to understand.

A fuzzy government can help in all three situations: it can be used to choose control parameters before starting the algorithm, to adjust them on-line to dynamically adapt to changing conditions, and to assist the user in assessing, designing, implementing, and validating an evolutionary algorithm for a given task. Table 7.1 shows two sets of rules worked out by Xu and colleagues to dynamically adjust the crossover and mutation rates of a genetic algorithm.

Herrera and Lozano in [89] describe an adaptive evolutionary algorithm where the fuzzy government adjusts two control parameters, $p_e$ and $\eta_{\min}$. The former governs the rate of application of an exploitation-oriented crossover operator, which is used along with a more conventional crossover operator with exploration properties: when $p_e$ is low, the conventional operator generates diversity and the algorithm takes up an exploration attitude; as $p_e$ increases, existing genetic material is exploited to generate better individuals. Parameter $\eta_{\min} \in [0, 1]$ determines selective pressure in a linear ranking selection mechanism (cf. chapter 1, section 1.7) $\eta_{\min}$ is inversely proportional to $\beta$). The joint management of these parameters allows the algorithm to strike a balance between exploration and exploitation. Genotypic diversity and phenotypic diversity are the inputs

| $\Delta p_e$ | Phenotypic Diversity | | |
|---|---|---|---|
| Genotypic Diversity | low | medium | high |
| low | medium | small | small |
| medium | large | large | medium |
| high | large | large | medium |

| $\Delta \eta_{\min}$ | Phenotypic Diversity | | |
|---|---|---|---|
| Genotypic Diversity | low | medium | high |
| low | small | medium | large |
| medium | small | large | large |
| high | small | small | large |

Table 7.2 Fuzzy rule-bases for the control of exploitation-oriented crossover rate and selective pressure respectively, suggested by Herrera and Lozano [89].

to this fuzzy government. The outputs are $\Delta p_e$ and $\Delta \eta_{\min}$, both comprised in $[\frac{1}{2}, \frac{3}{2}]$, which determine the change of $p_e$ and $\eta_{\min}$ as follows:

$$\begin{aligned} p_e &\leftarrow p_e \Delta p_e; \\ \eta_{\min} &\leftarrow \eta_{\min} \Delta \eta_{\min}. \end{aligned}$$

The fuzzy rule-bases for $\Delta p_e$ and $\Delta \eta_{\min}$ are shown in Table 7.2.

In [8], genetic programming is applied to symbolic regression and a fuzzy government is employed to dynamically adjust a number of parameters, such as the maximum length of genotypes, mutation probability, and stopping condition, and to detect the emergence of a solution.

A similar approach is adopted in [177], where an evolutionary algorithm for fuzzy controller synthesis and optimization, featuring a random tournament selection mechanism and noisy fitness, is in turn controlled by a fuzzy government, which exploits knowledge specific to controller optimization to direct search and make the best use of computational resources.

The random tournament, called *competition*, consists in testing two individuals (fuzzy logic controllers) on the same randomly generated initial state of the system to be controlled. A competition may have three outcomes:

1. either competitor violates a constraint (failure): the other wins;
2. either competitor succeeds in driving the system into a goal state (success): the other loses;

3. simulation times out, or both competitors fail or succeed at the same time (tie): the winner is decided at random.

Selective pressure can be adjusted by the use of multiple competition matches.

Available statistics are:

- genotypic diversity $D_\Gamma$, defined as the probability that two individuals randomly extracted from the population have different genotypes;
- phenotypic diversity $D_\Phi$, measuring the expected degree to which two individuals randomly extracted from the population have different fitness;
- maximum fitness $f^*$, average fitness $\bar{f}$, and minimum fitness $f_{\rm min}$;
- fitness range $R = f^* - f_{\rm min}$;
- tie rate $P_{\rm tie}$, defined as the fraction of random competitions that end in a tie;
- success rate $P_{\rm succ}$, i.e., the fraction of competitions that end with a success (even in the case of tie);
- actual mutation rate: this may differ from the mutation rate set through the relevant parameter, due to statistical fluctuations or failures to preserve the feasibility of a solution while attempting to mutate it;
- time-out rate $P_{\rm t/o}$, the fraction of competitions that time out.

The control parameters that can be acted upon are:

- mutation rate $p_m$;
- crossover rate $p_x$;
- selective pressure $S$;
- window of success $W$, i.e. the time a controller must spend in a goal state in order to be considered successful;

IF $P_{\text{tie}}$ is Z THEN $P_m$ is 0.0
IF NOT $P_{\text{tie}}$ is Z AND $P_{\text{tie}}$ is MS THEN $P_m$ is 0.004
IF NOT $P_{\text{tie}}$ is MS AND $P_{\text{tie}}$ is M THEN $P_m$ is 0.01
IF NOT $P_{\text{tie}}$ is M THEN $P_m$ is 0.1
IF $P_{\text{succ}}$ is S THEN $W$ is 0 ms
IF NOT $P_{\text{succ}}$ is S AND $P_{\text{succ}}$ is MS THEN $W$ is 10 ms
IF NOT $P_{\text{succ}}$ is MS THEN $W$ is 20 ms
IF VERY NOT $P_{\text{t/o}}$ is MS THEN Emergence is 0.5
IF NOT $P_{\text{t/o}}$ is VERY M THEN Emergence is 1.0
IF NOT $P_{\text{t/o}}$ is MS THEN $W$ is 20 ms

Figure 7.2 The rule set of a sample fuzzy government [177].

- emergence, that is the degree to which a satisfactory solution has been obtained;
- premature convergence, that is the degree to which evolution is stuck in a local optimum.

The latter two parameters are used to decide whether to stop the algorithm.

Figure 7.2 shows a sample fuzzy rule-base proposed by the authors to control the algorithm for an application to the ball-and-beam problem, a toy non-linear unstable control problem.

Since the aim of a fuzzy government is to keep the evolutionary process in a dynamical equilibrium, its overall structure will be based on negative feedback as a guiding principle, as it can be observed from the examples presented.

For instance, as a general rule, the mappings between population statistics and algorithm parameters defined by a fuzzy government will inversely relate the mutation rate with diversity (the less diverse the population, the higher the mutation rate).

Finally, it has to be remarked that statistics and control parameters are in part universal to any evolutionary algorithm and in part specific to a particular application. Therefore it is hard to come up with general rules to control the evolutionary process. These are to be discovered by the implementer by means of experiments and empirical observations.

Visualization is a central issue in studying the behavior of an evolutionary algorithm, mainly because of the huge quantity of data and the need to provide them in a concise, clear, and readable form. Besides, the fact should be stressed that almost everything one learns on how an evolutionary algorithm works, is learned through the inspection of the automatically generated execution reports of a large number of test runs. For this reason a comprehensive yet terse visualization and report generation is a vital step toward the implementation of a fuzzy government.

The quantity of data associated to each generation complicates the development of a good report-generating function. Because it is impossible to visualize all the data, a small but significant subset thereof must be selected, that captures the situation at a given time. The size of this data set is constrained by the need to keep the computational cost of the report-generating function below a small percentage of the overall cost.

### 7.2.4 Automatic Learning

An automatic learning process was proposed by Lee and Takagi [122, 125] to find out fuzzy rule bases along with the appropriate definition of the relevant membership functions. The mechanism bears many resemblances with Grefenstette's meta-GA [80]. A genetic algorithm runs on top of the evolutionary algorithm to be controlled, whose individuals encode possible fuzzy governments for the object evolutionary algorithm. The fitness function for this meta-GA is based on the on-line and off-line performance measures [108], obtained from the object evolutionary algorithm by applying it to the five DeJong's test functions.

As an initial demonstration of their method, Lee and Takagi used a fuzzy system taking three input variables (statistics) and producing three output variables (parameters). The three input variables are

- the ratio between the average fitness and the best fitness, $\bar{f}/f^*$;
- the ratio between the worst fitness and the average fitness, $f_{\min}/\bar{f}$;
- the change in fitness since the last control action.

The evolutionary algorithm parameters under control are population size, mutation rate, and crossover rate. These parameters are not acted upon directly; rather, the output variables of the fuzzy government are the changes in these parameters.

Experiments carried out by Lee and Takagi [122] showed that an evolutionary algorithm for the design of fuzzy controllers for the inverted pendulum using the fuzzy government obtained by their method exhibited much better performance than a simple static algorithm. Other experiments aimed at isolating the effects of the individual parameters on the overall algorithm performance suggested that the most important contribution comes from adapting the mutation rate [125].

### 7.2.5 Co-evolution with a Fuzzy Government

An alternative approach to writing a fuzzy government or learning it once and for all using a set of test problems as a benchmark for evolutionary algorithm performance is to have the fuzzy government co-evolve along with the solution to the object problem.

This idea is due to Herrera and Lozano [90], who call *Fuzzy Behaviors* the representation of rule consequents of the fuzzy government. During the application of the genetic operator being controlled, a random assignment is made between each member of the population of fuzzy behaviors and sets of parents. Then, the genetic operator is applied to each set using the control parameter value obtained by applying a fuzzy government using the rule represented by the corresponding fuzzy behavior.

The population of fuzzy behaviors will undergo evolution, through the effects of its own selection process as well as mutation and crossover. The fitness of the fuzzy behaviors will depend on the effectiveness of the genetic operators they controlled. Operator effectiveness may be measured according to different criteria, such as whether the operator generated offspring fitter than their parents, or introduced diversity in the population.

## 7.3 Evolutionary Algorithms with Fuzzy Components

### 7.3.1 Fuzzy Fitness

In some applications, it is hard to assess the quality of candidate solutions with precision. This is usually due to the fact that a solution should be tested against a huge or infinite number of cases and fitness would be a function of the scores obtained in each test. This is for example the case in genetic programming, where a sample of fitness cases is usually selected *a priori*, in the hope that it be representative of the whole gamut of possible inputs. In other types of applications, fitness is inherently noisy (i.e., its calculation yields different results each time it is performed).

In all cases, a better fitness estimate can be obtained at the cost of repeating its calculation as many times as desired. However, it might be argued that knowing the fitness of an individual beyond a certain precision is useless, given the random nature of evolutionary algorithms. In other words, what is the point of knowing the fitness of an individual up to 8 decimal digits if that fitness is used by a stochastic selection operator?

A sensible approach to spare computational resources is to treat fitness imprecision with the instruments provided by fuzzy set theory, like the ones illustrated in Chapter 3.

An example of how a fuzzy fitness may be defined is the following, based on a control problem, where pairs of controllers play competitions which can end up in a success of either competitor, in a tie, or in a failure of either competitor: in such setting, a fuzzy estimate of the fitness of individuals may be given based on the recorded outcomes of competitions. This estimate is based on three quantities: the number $c$ of competitions undergone by an individual, the number $w$ of its wins, and the number $s$ of successes. The more competitions an individual has gone through, the less fuzzy its fitness will be.

The membership function of fitness, for a given individual, is defined as:

$$\mu_f(x) = N(a,b)x^a(1-x)^b, \tag{7.1}$$

where

$$N(a,b) = \frac{(a+b)^{a+b}}{a^a b^b} \tag{7.2}$$

is a normalization factor and $a = w + s$, $b = c - s$. The mode of the membership function, where its value is one, is

$$\arg\max_{x\in[0,1]} \mu_f(x) = \frac{a}{a+b} = \frac{w+s}{w+c}. \tag{7.3}$$

An interesting predicate defined on pairs of individuals is their *fitness overlap:*

$$f_\wedge(\gamma,\kappa) = \max_{x\in[0,1]} \min\{\mu_{f(\gamma)}(x), \mu_{f(\kappa)}(x)\}, \tag{7.4}$$

i.e. the maximum of the intersection of their fitnesses. This gives the degree to which $\gamma$ and $\kappa$ have the same fitness.

The fuzzy fitness is employed to avoid carrying out useless competitions: when the fitness overlap of the two opponents is below a certain threshold, the one with the highest mode is awarded the match.

Fuzzy fitnesses of individuals in the population are aggregated to yield statistics such as minimum, average, and maximum fitness for use by the fuzzy government.

### 7.3.2 Recombination Operators Based on Fuzzy Connectives

The observation that recombination plays a central role in the exploration/exploitation balance, which is essential to evolutionary algorithms, brings forth the question whether it is possible to design recombination operators allowing this balance to be tuned.

At least for real-coded evolutionary algorithms, a solution consists in defining a recombination operator based on fuzzy connectives, those in turn based on triangular norms and co-norms and average functions.

Assume that $\mathbf{x} = (x_1, \ldots, x_N)$ and $\mathbf{y} = (y_1, \ldots, y_N)$ are two real-coded genomes to which recombination has to be applied. The gene in each locus $i$ may have alleles comprised within a given interval, $x_i, y_i \in [a_i, b_i] \subseteq \mathbb{R}$. Assuming without loss of generality that, for a given $i$, $x_i < y_i$, we observe that the space of alleles for that gene can be partitioned into three intervals: $[a_i, x_i]$, $[x_i, y_i]$, and $[y_i, b_i]$. The middle interval can be classified as an exploitation interval, whereas the external intervals can be thought of as exploration zones.

Based on this idea, three functions can be defined to be used by the recombination operator[92], that map the alleles in the two parents in the allele in a child, in such a way that the result falls respectively below, in the middle of, or above the parent alleles. Let us call these three families of functions $L$ (for "left", because its value is at the left of its two arguments), $M$ (for "middle"), and $R$ (for "right"). Formally, for all $x, y \in [a, b] \subseteq \mathbb{R}$,

$$L(x, y) \leq \min\{x, y\}, \tag{7.5}$$

$$\min\{x, y\} \leq M(x, y) \leq \max\{x, y\}, \tag{7.6}$$

$$\max\{x, y\} \leq R(x, y). \tag{7.7}$$

Now, it happens that $L$ functions have just the properties of triangular norms (cf. Chapter 3, Section 3.2.4), while $R$ functions have the properties of triangular co-norms. $M$ functions are so-called averaging functions.

Having defined these three families of functions, three recombination operators can be defined based on them: $L$-, $M$-, and $R$-recombination, featuring different exploration/exploitation behaviors. By mixing these three recombination operators with different (possibly dynamic) probabilities, one can achieve a fine control over the exploration/exploitation behavior of the whole evolutionary algorithm.

### 7.3.3 Soft Genetic Operators

*Soft modal* recombination and mutation operators for real-valued encodings, based on triangular probability distributions, were introduced by Hans-Michael Voigt and colleagues [231, 232, 234] as a generalization of the discrete crossover and mutation used in the *breeder* genetic algorithm (cf. chapter 1).

#### Soft Modal Recombination Operator

This recombination operator, is also called *fuzzy recombination* in [234] for being inspired by fuzzy set theory. Given two parent chromosomes $(x_1, \ldots, x_N)$ and $(y_1, \ldots, y_N)$, the probability that the offspring have value $z_i$, for $i = 1, \ldots, N$, has a bimodal distribution

$$p(z_i) \in \{\phi(x_i), \phi(y_i)\}, \tag{7.8}$$

where $\phi(r)$ is a triangular probability distribution with mode $r$ defined in $[r - d|y_i - x_i|, r + d|y_i - x_i|]$ as

$$\phi(z) = \begin{cases} \frac{z-r+d|y_i-x_i|}{d^2(y_i-x_i)^2}, & z \leq r, \\ \frac{r+d|y_i-x_i|-z}{d^2(y_i-x_i)^2}, & z > r, \end{cases} \tag{7.9}$$

with $d \geq 1/2$.

#### Soft Modal Mutation

Given a real-valued allele $x$ for a gene defined in $[a, b]$ to be mutated, soft modal mutation [232] generates a new allele $x' = x + \Delta$, where $\Delta$ has a distribution $p(\Delta)$ randomly chosen from the family

$$\{\phi(\pm A\beta^{\pi}), \phi(\pm A\beta^{\pi+1}), \ldots, \phi(\pm A\beta^{0})\}, \tag{7.10}$$

where $A \ll b - a$ is the amplitude of mutation, $\pi = \lfloor \log_\beta R_{\min} \rfloor < 0$, with $\beta > 1$ called the *base* of the mutation and $R_{\min}$ the lower limit of the relative mutation change and $\phi(x)$ defined in a way analogous to Equation 7.9.

### 7.3.4 Recombination Using Templates

When genotypes consist of strings of real numbers in the interval $[0, 1]$, representing fuzzy sets [199], a fuzzy recombination operator can be defined which makes use of templates. A genotype in binary-coded genetic algorithms can be represented with the set of all loci where the alleles are 1, for example a chromosome $\gamma = 011101$ would be described by set $S_\gamma = \{2, 3, 4, 6\}$. Recombination in binary-coded genetic algorithms could then be defined in terms of these sets of 1-loci: for instance, one-point crossover between $\gamma$ and $\kappa$ at position $i$, giving two offspring

$$\begin{aligned} \chi_1 &= \gamma_1 \cdots \gamma_i \kappa_{i+1} \cdots \kappa_N \\ \chi_2 &= \kappa_1 \cdots \kappa_i \gamma_{i+1} \cdots \gamma_N, \end{aligned}$$

might be described in terms of the associated sets $S_{\chi_1}$ and $S_{\chi_2}$, by means of the set $T = \{i + 1, \ldots, N\}$ of loci where material is exchanged between the parents, the recombination *template*, as follows:

$$\begin{aligned} S_{\chi_1} &= (S_\gamma \cap \bar{T}) \cup (S_\kappa \cap T), \\ S_{\chi_2} &= (S_\gamma \cap T) \cup (S_\kappa \cap \bar{T}), \end{aligned} \tag{7.11}$$

where $\bar{T}$ is the complement of $T$ in the universe of all possible loci.

Having described a recombination operator this way, one can generalize by letting the template $T$ be a fuzzy set. Assuming the alleles of each single gene are degrees of membership or truth values, this generalization is straightforward if one considers that the $\cap$ and $\cup$ logical connectives in Equation 7.11 are the fuzzy ones.

### 7.3.5 Representation

The recent trend in evolutionary algorithms has been toward a more complex mapping between the genotype and the phenotype, reminiscent of the ontogeny as opposed to phylogeny in natural genetics, i.e., the development of the phenotype of a single individual in response to the environment from its embryo, which is uniquely determined by the genotype.

An idea that goes along these lines consists in defining a *fuzzy representation* whereby each feature of a solution to the object problem is determined by a number of associated fuzzy decision variables in $[0, 1]$.

As an instance of this type of representation [230], when tackling continuous parameter optimization, each problem parameter might have $m$ fuzzy decision variables associated with it. For each parameter, the decoding is carried out by means of a function $g : [0, 1]^m \to [0, 1]$ and a linear transformation from the interval $[0, 1]$ to the relevant parameter domain $[a, b]$. An example of $g(\cdot)$ is [233]

$$g(\mathbf{d}) = \frac{1}{2^{m-1} - 1} \sum j = 1^m d_j 2^{j-1}, \qquad (7.12)$$

where $\mathbf{d}$ is the vector of $m$ fuzzy decision variables.

When $m > 1$, this type of coding breaks the one-to-one correspondence between genotype and phenotype. Along with a fuzzy representation, genetic operators based on fuzzy logic can be defined [230].

Experimental evidence [233] suggests that the use of a fuzzy representation allows a robust behavior to be obtained, and a better performance for small-sized populations.

Chapter 8

# Natural Parallel (Soft) Computing

## 8.1 Introduction

MANY problem solving and heuristic techniques, including those typical of soft computing, have the distinctive feature of being directly or indirectly inspired by the observation of the natural world. If we look, for instance, into such processes as biological evolution or the functioning of the brain, we notice that many things are happening at the same time. The same can be said of many other natural systems such as insect societies and ecologies in general. In other words, these systems are natural massively parallel ones where more or less simple agents, such as nervous cells or ants, work jointly, in a distributed manner, to sustain the whole or to "solve" a problem. In short, many, if not all, natural systems work on a problem in a collective, concerted manner. Of course, collectively "solving" a problem does not have the same meaning in nature as in the sciences. Solving a problem might mean building a bee nest, firing a few million interconnected neurons in response to a stimulus, or just surviving in an animal ecology. There is no explicit concept of "optimizing" something, nor of finding a particular solution, just of adapting so as to maintain collective viability in the face of a changing environment. This is in the spirit of soft computing where approximation, adaptivity, redundancy, and fault tolerance are encouraged in order to obtain at least robust, if not optimal, solutions, apart from the fact that for man-made systems we often know what good solutions should look like. These parallel and distributed interacting processes are interesting in themselves and may suggest new powerful ideas and paradigms for problem solving. If time is deemed important, as is the case for many difficult search and optimization problems, parallelism also offers the possibility of efficiency gains by sharing the load over many agents or workers.

In contrast, our general-purpose problem-solving machine, the electronic computer, is built in a manner that reflects historical developments and technological and economical constraints. Essentially, the standard computer is a sequential machine in which a stream of instructions is executed on a stream of data, both stored in memory. Clearly, this is only a very schematic picture. Today's processors and memories have a great deal of built-in parallelism and a number of processors can be pieced together via some interconnection to function in parallel. Later we will present the fundamental architectural aspect of parallel and distributed computer systems and their use in soft computing methodologies. For the time being, however, it should be noted that the apparent mismatch between the artificial counterparts of natural parallel systems and the standard computer is not that important in principle, due to the simulation capabilities of the machine. In other words, even without a strict match between the distributed algorithmic schema and the machine, and if we make abstraction of computing time, most interesting distributed systems can be simulated by suitable interleavings on a serial machine. This in turn means that we can focus our attention on the architectural parallel and distributed aspects of soft computing algorithms, without being too concerned with implementations at the model level. Of course, actual parallel implementation models are important in practice and will be described in detail.

The next section is a short introduction to the main types of parallel and distributed architectures that can be used in soft computing applications. In the following sections we will examine the intrinsic parallel and distributed features of two of the three main soft computing approaches presented in this book, namely evolutionary algorithms and neural networks. We will study their general architectural and dynamical aspects, describing practical implementations on real hardware as needed. Fuzzy systems are less convincing as natural parallel systems and have not been included.

## 8.2 Parallel and Distributed Computer Architectures: An Overview

PARALLELISM can arise at a number of levels in computer systems: task level, instruction level, or at some lower machine level. Although there are several ways in which parallel architectures may be classified, the standard model of Flynn is widely accepted as a starting point. But the reader should be aware that it is a coarse-grain classification: for instance, even today's serial processors are in fact parallel in the way in which they execute instructions, as well as with respect to the memory interface. Even at a higher architectural level, many parallel architectures are in fact hybrids of the base classes. For a comprehensive treatement of the subject, the reader is referred to Hwang's text [101].

Flynn's taxonomy is based on the notion of instruction and data streams. There are four possible combinations conventionally called *SISD* (single instruction, single data stream), *SIMD* (single instruction, multiple data stream), *MISD* (multiple instruction, single data stream), and *MIMD* (multiple instruction, multiple data stream). Figure 8.1 schematically depicts the three most important model architectures.

The SISD architecture corresponds to the classical mono-processor machine such as the typical PC or workstation. As stated above, there is already a great deal of parallelism in this architecture at the instruction execution level (pipelining, superscalar and very long instruction word). This kind of parallelism is "invisible" to the programmer in the sense that it is built-in in the hardware or must be exploited by the compiler.

In the SIMD architecture the same instruction is broadcast to all processors. The processors operate in lockstep executing the given instruction on different data stored in their local memories (hence the name: single instruction, multiple data). The processor can also remain idle, if this is appropriate. SIMD machines exploit *spatial* parallelism that may be present in a given problem and are suitable for large, regular data structures. If the problem domain is spatially or temporally irregular, many processors must remain idle at a given time step since different operations are called for in different regions. Obviously, this entails a serious loss in the amount of parallelism that can be exploited. Another type of SIMD computer is the vector processor which is specialized in the pipelined execution of numerical array operations and is the typical component of supercomputers.

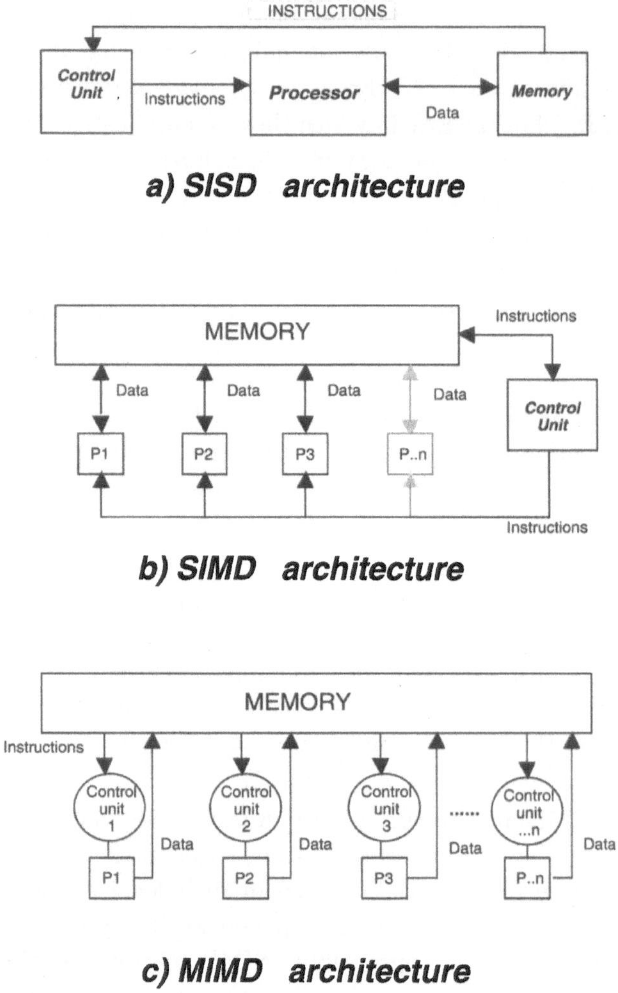


Figure 8.1 Flynn's model of parallel architectures. The shared memory model is depicted; in the SIMD and MIMD case memory can also be distributed (see text).

In the MIMD class of parallel architectures multiple processors work together through some form of interconnection. Different programs and data can be loaded into different processors which means that each processor can execute different instructions at any given point in time. Of course, usually the processors will require some form of synchronization and communication in order to cooperate on a given application. This class is the more generally useful and most commercial parallel and distributed architectures belong to it.

There has been little interest up to now in the MISD class since it does not lend itself to readily exploitable programming constructs. One type of architecture of this class that enjoys some popularity are the so-called *systolic arrays* which are used in specialized applications such as signal processing.

Another important design decision is whether the system memory spans a single address space or it is distributed into separated chunks that are addressed independently. The first type is called *shared memory* while the latter is known as *distributed memory*. This is only a logical subdivision independent of how the memory is physically built.

In shared memory multiprocessors all the data are accessible by all the processors. This poses some design problems for data integrity and for efficiency. Fast cache memories next to the processors are used in order to speedup memory access to often-used items. Cache coherency protocols are then needed to insure that all processors see the same value for a given piece of data.

Distributed memory multicomputers is also a popular architecture which is well suited to most parallel workloads. Since the address spaces of each processor are separate, communication between processors must be implemented through some form of message passing. To this class belong networked computers, sometimes called computer *clusters*. This kind of architecture is interesting for several reasons. First of all, it has a low cost since already existing local networks of workstations can be used just by adding a layer of communication software to implement message passing. Second, the machines in these networks usually feature up-to-date off-the-shelf processor and standard software environments which make program development easier. The drawbacks are that parallel computing performance is limited by comparatively high communication latencies and by the fact that the machines have different workloads at any given time and are possibly heterogeneous. Nevertheless, problems that do not need frequent communication are suitable for this architecture. Moreover, some of the drawbacks can be overcome by using networked computers in dedicated mode with a high-performance communication switch. Although this solution is more expensive, it can be cost-effective with respect to specialized parallel machines.

Finally, one should not forget that the World Wibe Web provides important infrastructures for distributed computation. As it implements a general distributed computing model, this Web technology can be used for parallel computing and for both computing and information related applications. Harnessing the Web or some other geographically distributed computer resource so that it looks like a single computer to the user is called *metacomputing* . The concept is very attractive but many challenges remain. In fact, in order to transparently and efficiently distribute a given computational problem over the available resources without the user taking notice requires advances in the field of user interfaces and in standardized languages, monitoring tools, and protocols to cope with the problem of computer heterogeneity and uneven network load. The Java environment is an important step in this direction.

We will see in the following sections how the different architectures can be used for distributed evolutionary computing, neural networks, and fuzzy systems.

## 8.3 Parallel and Distributed Evolutionary Algorithms

WE SAW in Chapter 1 how evolutionary algorithms find their inspiration in the evolutionary processes occurring in nature. For historical and simplicity reasons, the standard description of EAs is sequential (see Chapter 1, Section 1.2). This is a perfectly reasonable and understandable starting point. However, it is too tied to the artificial constraints of sequential computer architecture, which was in practice the only one widely available at EA's inception time. This view is limited as it tends to conceal some EA's important aspects. Actually, if one were to follow more faithfully the biological inspiration, parallel and distributed EAs should be the norm, rather than the exception. Today, many parallel and distributed systems are readily available and even sequential machines have reached such a power that simulations of distributed processes on serial hardware can be practical. As well, there is another reason why parallel algorithms are attractive: the need for increased performance. Even though processors are more powerful each year, when applied to large hard problems evolutionary algorithms are comparatively slow. One way to overcome time and size constraints in a computation is by adding processors, memory, and an interconnection network and putting them to work together on a given problem. By sharing the workload, it is hoped that an $N$-processor system will give rise to a *speedup* in the computation time. Speedup is defined as the the time it takes a single processor to execute a given problem instance with a given algorithm divided by the time on an $N$-processor architecture of the same type for the same problem instance and with the same algorithm. Sometimes, a different, more suitable algorithm is used in the parallel case. Clearly, in the ideal case the maximum speedup is equal to $N$. If speedup is indeed almost linear in the number of processors, then time consuming problems can be solved in parallel in a fraction of the uniprocessor time or larger and more interesting problem instances can be tackled in the same amount of time. Actually, things are not so simple since in most cases several overhead factors contribute to significantly lower the theoretical performance improvement. Furthermore, general parallel programming models turn out to be difficult to design due to the large architectural space they must span and to the resistance represented by current programming paradigms and languages. In any event, many important problems are sufficiently regular in their space and time dimensions as to be suitable for parallel or distributed computing and evolutionary algorithms are certainly among those.

### 8.3.1 Parallel and Distributed Evolutionary Algorithms Models

There are two main reasons for parallelizing an evolutionary algorithm: one is to achieve time savings by distributing the computational effort and the second is to benefit from a parallel setting from the algorithmic point of view, in analogy with the natural parallel evolution of spatially distributed populations.

A first type of parallel evolutionary algorithm makes use of the available processors or machines to run independent problems. This is trivial, as there is no communication between the different processes and for this reason it is sometimes called a *pleasingly parallel* algorithm (it used to be called "embarassingly parallel"). This extremely simple method of doing simultaneous work can be very useful. For example, this setting can be used to run several versions of the same problem with different initial conditions, thus allowing gathering statistics on the problem. Since evolutionary algorithms are stochastic in nature, being able to collect this kind of statistics is very important. This method is in general to be preferred with respect to a very long single run since improvements are more difficult at later stages of the simulated evolution. Other ways in which the model can be used is to solve $N$ different versions of the same problem or to run $N$ copies of the same problem but with different GA parameters, such as crossover or mutation rates. Neither of the above adds anything new to the nature of the evolutionary algorithms but the time savings can be large.

We now turn to genuine parallel evolutionary algorithm models. There are several possible levels at which an evolutionary algorithm can be parallelized: the population level, the individual level or the fitness evaluation level. Moreover, although genetic algorithms and genetic programming are similar in many respects, the differences in the individual representation make genetic programming a little bit different when implemented in parallel. The next section describes the parallelization of the fitness evaluation while the two following sections treat the population and the individual cases respectively.

### 8.3.2 Global Parallel Evolutionary Algorithms

Parallelization at the fitness evaluation level does not require any change to the standard evolutionary algorithm since the fitness of an individual is independent of the rest of the population. Moreover, in many real-world problems, the calculation of the individual's fitness is by far the most time consuming step of the algorithm. This is also a necessary condition in order for the communication time to be small in comparison to the time spent in computations. In this case an obvious approach is to evaluate each individual fitness simultaneously on a different processor. A *master* process manages the population and hands out individuals to evaluate to a number of *slave* processes. After the evaluation, the master collects the results and applies the genetic operators to produce the next generations. Figure 8.2 graphically depicts this architecture. If there are more individuals than processors, which is often the case, then the individuals to be evaluated are divided as evenly as possible among the available processors. This architecture can be implemented on both shared memory multiprocessors as well as distributed memory machines, including networked computers. For genetic algorithms it is assumed that fitness evaluation takes about the same time for any individual. The other parts of the algorithm are the same as in the sequential case and remain centralized. The following is an informal description of the algorithm:

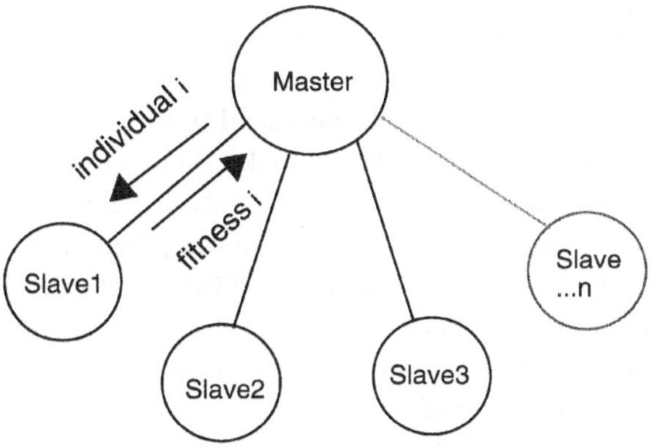


Figure 8.2 A schematic view of the master-slave model.

Produce an initial population of individuals
**for** all individuals **do in parallel**
  Evaluate the individual's fitness
**end parallel for**
**while not** *termination condition* **do**
  Select fitter individuals for reproduction
  Produce new individuals
  Mutate some individuals
  **for** all individuals **do in parallel**
    Evaluate the individual's fitness
  **end parallel for**
  Generate a new population by inserting some new good
  individuals and by discarding some old bad individuals
**end while**

In the GP case, due to the widely different sizes and complexities of the individual programs in the population different individuals may require different times to evaluate. This in turn may cause a load imbalance which decreases the utilization of the processors. Oussaidène *et al.* [161] implemented a simple method for load-balancing the system on a distributed memory parallel machine and observed nearly linear speedup. However, load balancing can be obtained for free if one gives up generational replacement and permits steady-state reproduction instead see Chapter 1, Section 1.7.1 and Section 8.3.5).

### 8.3.3 Island Distributed Evolutionary Algorithms

We now turn to individual or population-based parallel approaches for evolutionary algorithms. All these find their inspiration in the observation that natural populations tend to possess a *spatial* structure. As a result, so-called *demes* make their appearance. Demes are semi-independent groups of individuals or subpopulations having only a loose coupling to other neighboring demes. This coupling takes the form of the slow migration or diffusion of some individuals from one deme to another. A number of models based on spatial structure have been proposed. The two most important categories are the *island* and the *grid* models.

The *island* model [47] features geographically separated subpopulations of relatively large size. Subpopulations may exchange information from time to time by allowing some individuals to migrate from one subpopulation to another according to various patterns. The main reason for this approach is to periodically reinject diversity into otherwise converging subpopulations. As well, it is hoped that to some extent, different subpopulations will tend to explore different portions of the search space. When the migration takes place between nearest neighbor subpopulations the model is called *stepping stone*. Within each subpopulation a standard sequential evolutionary algorithm is executed between migration phases. Figure 8.3 depicts this distributed model and the following is a schematic algorithmic description of the process:

```
Initialize P subpopulations of size N each
generation = 0
while not termination condition do
  for each subpopulation do in parallel
    Evaluate and select individuals by fitness
    if generation mod frequency = 0 then
      Send K < N best individuals to
      a neighboring subpopulation
      Receive K individuals from a
      neighboring population
      Replace K individuals in
      the subpopulation
    end if
    Produce new individuals
    Mutate individuals
  end parallel for
  generation = generation + 1
end while
```

Here *frequency* is the number of generations before an exchange takes place. Several individual replacement policies have been described in the literature. One of the most common is for the migrating K individuals to displace the K worst individuals in the subpopulation. It is to be noted that the subpopulation size, the frequency of exchange, the number of migrating individuals, and the migration topology are all new parameters of the algorithm that have to be set in some way. At present there is no rigorous way for choosing them. However, several works have empirically arrived at rather similar parameter values [211, 47]. Several migration topologies have been used: one is the "ring", in which populations are topologically disposed along a circle and exchanges take place between neighboring subpopulations, as illustrated in Figure 8.4. Another possibility are "meshes of islands", possibly toroidally interconnected, in which migration is towards nearest neighbor nodes (also shown in Figure 8.4). One possible drawback of these *static* topologies is that some bias might be introduced by the repeating exchange pattern. *Dynamical* topologies, where destination nodes change during time, seem more useful for preserving diversity in the subpopulation. For example, good results have been obtained with the use of a "random" topology [62] . This can be implemented by sending groups of individuals from the islands to a central buffer from which each node takes its "immigrants" as needed. In this scheme, it is easy to take care of limiting conditions such as a node sending to itself or being chosen too often as a target.

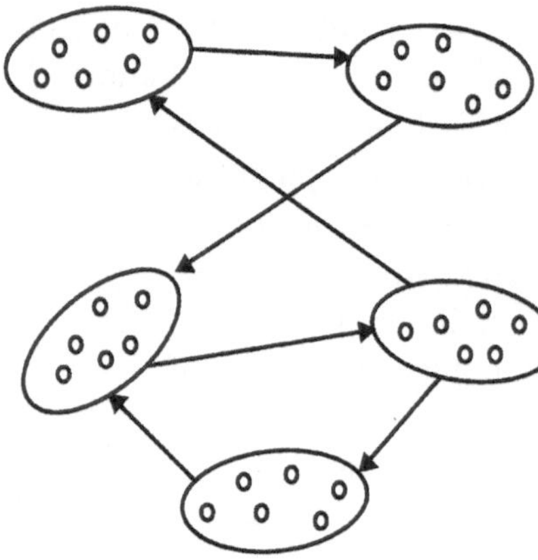

Figure 8.3 The *island* model of semi-isolated populations.

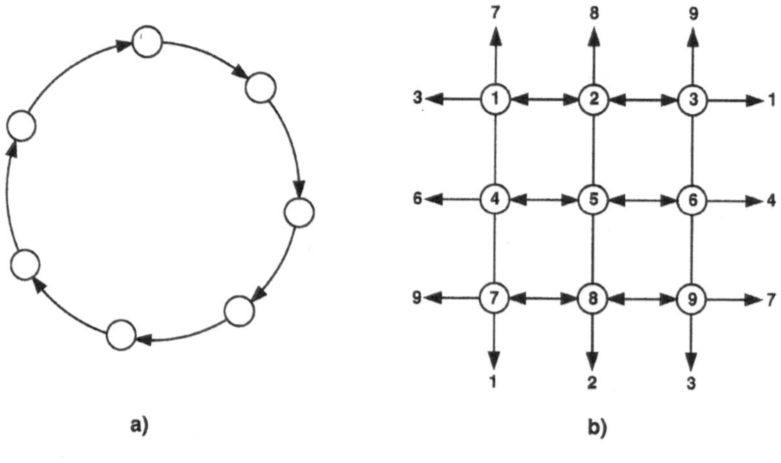


Figure 8.4 Two commonly used distributed EA topologies: a) the "ring" topology, b) the "mesh" topology. In the mesh topology processors "wrap around", giving a toroidal structure, as shown by the numbered arrows.

### 8.3.4 Cellular Genetic Algorithms

In the *grid* or *fine-grained* model [133] individuals are placed on a large toroidal (the ends wrap around) one, two, or three-dimensional grid, one individual per grid location (see Figure 8.5). The model is also called *cellular* because of its similarity with cellular automata with stochastic transition rules [222, 240].

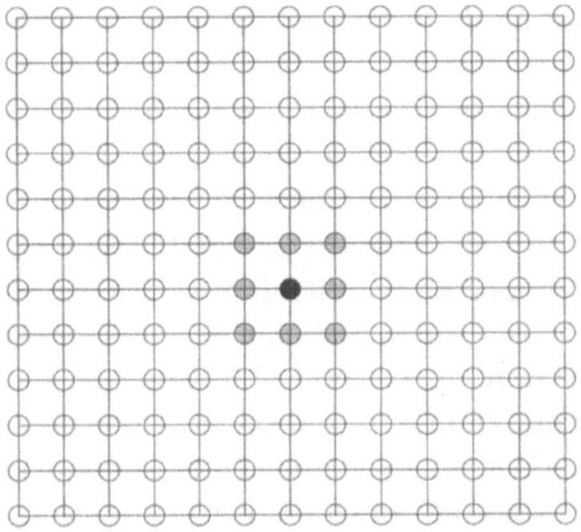

Figure 8.5 A 2-D spatially extended population of individuals. A possible neighborhood of an individual (black) is marked in gray.

Fitness evaluation is done simultaneously for all individuals and selection, reproduction, and recombination take place locally within a small neighborhood. In time, semi-isolated niches of genetically homogeneous individuals emerge across the grid as a result of slow individual diffusion. This phenomenon is called *isolation by distance* and is due to the fact that the probability of interaction of two individuals is a fast-decaying function of their distance. The following is a pseudo-code description of a grid evolutionary algorithm.

**for** each cell $i$ in the grid **do in parallel**
    generate a random individual $i$
**end parallel for**
**while not** *termination condition* **do**
    **for** each cell $i$ **do in parallel**
        Evaluate individual $i$
        Select a neighboring individual $k$
        Produce offspring from $i$ and $k$
        Assign one of the offspring to $i$
        Mutate $i$
    **end parallel for**
**end while**

In the preceding description the neighborhood is generally formed by the four or eight nearest neighbors of a given grid point (see Figure 8.5). In the 1-D case a small number of cells on either side of the central one is taken into account. The selection of an individual in the neighborhood for mating with the central individual can be done in various ways. Tournament selection is commonly used since it matches nicely the spatial nature of the system. Local tournament selection extracts $k$ individuals from the population with uniform probability but without re-insertion and makes them play a "tournament", which is won, in the deterministic case, by the fittest individual among the participants. The tournament may be probabilistic as well, in which case the probability for an individual to win it is generally proportional to its fitness. This makes full use of the available parallelism and is probably more appropriate if the biological methaphor is to be followed. Likewise, the replacement of the original individual can be done in several ways. For example, it can be replaced by the best among itself and the offspring or one of the offspring can replace it at random. The model can be made more dynamical by adding provisions for longer range individual movement through random walks, instead of having individuals interacting exclusively with their nearest neighbors [222]. A noteworthy variant of the cellular model is the *cellular programming algorithm* [210]. Cellular programming has been extensively used to evolve cellular automata for performing computational tasks.

### 8.3.5 Implementation Aspects and Experimental Observations on Parallel EAs

The master-slave global fitness parallelization model can be implemented on both shared memory multiprocessors and distributed memory multicomputers, including workstation clusters, and it is well adapted when fitness calculation takes up most of the computation time. However, it is well known that distributed memory machines have better scaling behavior. This means that performance is relatively unaffected if more processors (and memory) are added and larger problem instances are tackled. In this way, computation and communication times remain well balanced. Shared memory machines, on the other hand, suffer from memory access contention, especially if global memory access is through a single path such as a bus. Cache memory alleviates the problem but it does not solve it.

Although both island and cellular models can be simulated on serial machines, thus comprising useful variants of the standard globally communicating GA, they are ideally suited for parallel computers. From an implementation point of view, coarse-grained island models, where the ratio of computation to communication is high, are more adapted to distributed memory multicomputers and for clusters of workstations. Recently, some attempts have been made at using Web and Java-based computing for distributed evolutionary algorithms [154]. The advantages and the drawbacks of this approach have been discusses in Section 8.2.

Massively parallel SIMD (Single Instruction Multiple Data) machines such as the Connection Machine of the late 1980s are appropriate for cellular models, since the necessary local communication operations, though frequent, are very efficiently implemented in hardware.

Parallel and distributed genetic programming has usually been modeled as a number of interacting populations, that is according to the island model. Parallel GP has also been successfully modeled as a fine-grained population where each individual occupies one cell of the grid [72]. This is perfectly possible if the topology is simply simulated on a sequential computer. However, since individuals in GP may widely vary in size and complexity cellular implementations of GP on SIMD arrays is difficult both because of the amount of local memory needed to store individuals as well as for efficiency reasons (e.g., sequential execution of different branches of code belonging to individuals stored on different processors). A more efficient parallel or distributed implementation of cellular GP would partition the whole population into subpopulations where each subpopulation is a grid of individuals. This model could easily be implemented on a distributed memory machine or a cluster. SIMD machines can be used with genetic programming in the fitness evaluation phase. Each processor may have the same individual evaluated on a *different* fitness case. Subsequently, the total fitness of the individual is obtained by a parallel sum of the single fitness cases.

In general, it has been found experimentally that parallel genetic algorithms, apart from being significantly faster, may help in alleviating the premature convergence problem and are effective for multimodal optimization. This is due to the larger total population size and to the relative isolation of the spatial regions where solutions start to co-evolve. Both of these factors help to preserve diversity while at the same time promoting local search. Several results in the literature support these conclusions, see for instance [126, 129, 211, 152]. As in the sequential case, the effectiveness of the search can be improved by permitting hill-climbing, i.e. local improvement around promising search points [152]. It has even been reported that for the island model [126, 3] *superlinear* speedup has been achieved in some cases. Superlinear speedup means getting more than $N$-fold acceleration with $N$ processors with respect to the uniprocessor time. While superlinear speedup is strictly impossible for deterministic algorithms, it becomes possible when there is a random element in the ordering of the choices made by an algorithm. This has been shown to be the case in graph and tree search problems where, depending on the solution and the way in which the search space is subdivided, parallel search may be more than $N$-fold effective. The same phenomenon may occur in all kind of stochastic, Monte-Carlo type algorithms, including evolutionary algorithms. The net result is that in the multi-population case oftentimes *fewer* evaluations are needed to get the same solution quality than for the single population case with the same total number of individuals.

For coarse-grain, island-based parallel genetic programming the situation is somewhat controversial. Andre and Koza [3] reported excellent results on a GP problem implemented on a transputer network (the 5-parity problem). More recently, W. Punch [182] performed extensive experiments on two other problems: the "ant" problem and the "royal tree" problem. The first is a standard machine learning test problem [117], while the latter is a contructive benchmark for GP. Punch found that the multiple-population approach did not help in solving those problems. Although the two chosen problems were purposedly difficult for distributed GP approaches, these results point to the fact that the parallel, multi-population GP dynamics is not yet well understood. A recent study by Fernández *et al.* [63] has empirically shown that, in fact, depending on the problem, and using a random communication topology, there is an interval of parameter values such as subpopulation size and frequency of exchange that is in some sense "efficient" (it cannot be called optimal) in solving the given problem. Here efficiency is measured as the total computational effort in the sequential case compared to the distributed one. In spite of these empirical advances, it is fair to acknowledge that the subject is still a rather unsettled one, which is not surprising given that there is no solid theoretical basis yet even for the standard sequential GP.

Until now only the spatial dimension entered into the picture. If we take into account the temporal dimension as well, we observe that parallel evolutionary algorithms can be *synchronous* or *asynchronous*. Island models are in general synchronous, using SPMD

(Single Program Multiple Data), coarse-grain parallelism in which communication phases synchronize processes. This is not necessary and experiments have been carried out with asynchronous EAs in which subpopulations evolve at their own pace and exchange individuals only when some internally measured level of convergence has been attained. This avoids constraining all co-evolving populations to swap individuals at the same time irrespective of subpopulation evolution and increases population diversity. Asynchronous behavior can be easily and conveniently obtained in island models by using *steady-state* reproduction instead of generational reproduction, although generational models can also be made to work asynchronously. In steady-state reproduction a small number of offspring replace other members of the population as soon as they are produced instead of replacing all the population at once after all the individuals have gone through a whole evaluation-selection-recombination mutation cycle[143]. For genetic programming the asynchronous setting also helps for the load-balancing problem caused by the variable size and complexity of GP programs. By independently and asynchronously evaluating each program, different nodes can proceed at different speeds without any need for synchronization [3]. Asynchronous, distributed models are probably the most natural ones for artificially evolving populations.

Figure 8.6 schematically depicts the main parallel evolutionary algorithm classes according to their space and time dimensions. The reader should be aware that this is only a broad classification. Although the main classes are covered, a number of particular and hybrid cases exist in practice, some of which will be described in the next section.

### 8.3.6 Non-standard Parallel EA Models

Parallel evolutionary algorithms have been proposed that do not fit into any of the classes that have been described in the previous sections. From the topological and spatial point of view, the methods of parallelization can be combined to give *hybrid* models. Two-level hybrid algorithms are sometimes used. The island model can be used at the higher level, while another fine-grained model can be combined with it at the lower level. for example, one might consider an island model in which each island is structured as a grid

| | Coarse-grain | Fine-grain | |
|---|---|---|---|
| | Population | Individual | Fitness |
| Synchronous | **Island**<br>GA and GP | **Cellular**<br>GA<br>Synchronous, stochastic CA | **Master-slave**<br>GA and GP |
| Asynchronous | **Island**<br>GA and GP | asynchronous, stochastic CA | **Master-slave**<br>GA and GP |

Figure 8.6 Parallel and distributed algorithms classes according to their space and time behavior.

of individuals interacting locally. Another possibility is to have an island model at the higher level where each island is a global parallel EA in which fitness calculation is parallelized. Other combinations are also possible and a more detailed description can be found in [35].

Although these combinations may give rise to interesting and efficient new algorithms, they have the drawback that even more new parameters have to be introduced to account for the more complex topological structure.

So far, we have made the implicit hypothesis that the genetic material as well as the evolutionary conditions such as selection and crossover methods were the same for the different subpopulations in a multi-population algorithm. If one gives up these constraints and allows different subpopulations to evolve with different parameters and/or with different individual representations for the same problem then new distributed algorithms may arise. We will name these algorithms *non-uniform* parallel EAs. One recent example of this class is the *Injection island GA* (iiGA) of Lin *et al.* [126]. In an iiGA there are multiple populations that encode the same problem using a different representation size and thus different resolutions in different islands. The migration rules are also special in that migration is only one-way, going from a low-resolution to a high-resolution node.

According to Lin *et al.*, such a hierarchy has a number of advantages with respect to a standard island algorithm.

In biological terms, having different semi-isolated populations in which evolution takes place at different stages and with different, but compatible genetic material makes sense: it might also be advantageous for evolutionary algorithms. Genetic programming in particular might benefit from these ideas since the choice of the function and terminal sets is often a fuzzy and rather subjective issue.

### 8.3.7 Open Questions and Summary

Parallel evolutionary algorithms are more difficult to analyze than the sequential versions because they introduce a number of new parameters such as migration rates and frequencies, subpopulation size and topology, grid neighborhood shape and size among others. As a consequence, there seems to be lacking a better understanding of their working. Nevertheless, performance evaluation models have been built for the simpler case of the master-slave parallel fitness evaluation in the GA case [37] and for GP [161]. As well, some limiting simplified cases of the island approach have been modeled: a set of isolated demes and a set of fully connected demes [36]. These are only first steps towords establishing more principled ways for choosing suitable parameters for parallel EAs.

In another study, Whitley *et al.* [243] have presented an abstract model and made experiments of when one might expect the island model to out-perform single population GAs on separable problems. Linear separable problems are those problems in which the evaluation function can be expressed as a sum of independent nonlinear contributions. Although the results are not clear-cut, as the authors themselves acknowledge, there are indications that partitioning the population may be advantageous.

There have also been some attempts at studying the behavior of cellular genetic algorithms. For instance, Sarma and De Jong [201] investigated the effects of the size and shape of the cellular neighborhood on the selection mechanism. They found that the size of the neighborhood is a parameter that can be used to control the selection pressure over the population. In another paper [194] the convergence properties of a typical cellular algorithm were studied using the formalism of probabilistic cellular automata and Markov chains. The authors reached some conclusions concerning global convergence and the influence of the neighborhood size on the quality and speed at which solutions can be found. Recently, Capcarrère *et al.* [38] presented a number of statistical measures that can be applied to cellular algorithms in order to understand their dynamics. The proposed statistics were demonstrated on the specific example of the evolution of non-uniform cellular automata but they can be applied generally.

In conclusion, if one is to follow the biological methaphor, parallel and distributed evolutionary algorithms seem more natural than their sequential counterparts. When implemented on parallel computer architectures they offer the advantage of increased performance due to the execution of simultaneous work on different machines. Moreover, genuine evolutionary parallel models such as the island or the cellular model constitute new useful varieties of evolutionary algorithms and there is experimental evidence that they can be more effective as problem solvers. This may be due to several factors such as independent and simultaneous exploration of different portions of the search space, larger total population sizes, and the possibility of delaying the uniformization of a population by migration and diffusion of individuals. These claims have not been totally and unambiguously confirmed by the few theoretical studies that have been carried out to date. It is also unclear what kind of problem can benefit more from a parallel/distributed setting, although separable and multimodal problems seem good candidates.

One drawback of parallel and evolutionary algorithms is that they are more complicated than the sequential versions. Since a number of new parameters must be introduced such as communication topologies and migration/diffusion policies, the mathematical modeling of the dynamics of these algorithms is very difficult. Only a few studies have been performed to date and even those have been mostly applied to simplified, more tractable models. Clearly, a large amount of work remains to be done on these issues before a clearer picture can begin to emerge.

The effectiveness of parallel and distributed EAs has also been shown in practice in a number of industrial and commercial applications. Some real-life problems may need days or weeks of computing time to solve on serial machines. Although the intrinsic time complexity of a problem cannot be lowered by using a finite amount of computing resources in parallel, the use of parallelism often permits the reduction of these times to reasonable amounts. This can be very important in an industrial or commercial setting where the time to solution is instrumental for decision making and competitivity. Parallel and distributed evolutionary algorithms have been used successfully in operations research, engineering and manufacturing, finance, VLSI and telecommunication problems among others. The following section describes one such application in more detail.

## 8.4 A Case Study: Distributed GP for Trading Model Inference

It is commonly held that the behavior of financial markets is very difficult to forecast. Indeed, according to a widely accepted view, financial markets are unpredictable by their very nature since the price variations follow a random walk and any new information available to the traders is immediately reflected on the price and cannot be made use of. Nevertheless, many studies have shown that the distribution of price variations is non-Gaussian and that the random walk hypothesis is probably incorrect, or at least insufficient to explain the actual behavior of markets (see also Chapter 2, Section 2.5.4). Although the predictability of markets remains a dubious issue, these findings open the way to the interpretation of empirical statistical regularities in prices. Many techniques have been devised to make intelligent use of the abundant market data that are available. One group of popular techniques is called *technical analysis*. Technical analysis is an ensemble of methods that aim at the prediction of future asset prices by looking at past price time series. A *Trading model* is a sophisticated version of technical analysis in which trading rules based on past price data automatically generate trading recommendations for the user. For the sake of the present application, possible recommendations are $+1$, $-1$, or 0. The value $+1$ corresponds to a "buy signal", $-1$ corresponds to a "sell signal" and 0 corresponds to the neutral position (no exposure). Trading rules make use of so-called *indicators* which are functions of the price history, such as various kinds of moving averages. Thus, a trading model consists of a set of indicator computations and a collection of rules combining those indicators in various ways.

As an example, a simple trading model based on the difference of the price $x$ and a moving average $MA_x$ of the price could be of the form

$$g(I_x) = \text{sign}(I_x)\ f(|I_x|)$$

where $I_x$ is an indicator defined as

$$I_x = x - \text{MA}_x(\tau)$$

and

$$f(|I_x|) = \begin{cases} 1 & if \quad |I_x| > a \\ 0 & if \quad |I_x| < a \end{cases}$$

where $a > 0$ is a threshold constant and $\tau$ is the range of the moving average.

The basic idea in the application of GP to trading model induction is that the latter, which are essentially decision trees, can be seen as programs written with a limited set of functions and terminals. Thus, populations of trading models can be made to evolve and automatically learn complex trading decision rules, given an adequate fitness measure. A complete description is to be found in [46], here we will give a summary of that work.

### 8.4.1 Indicators and Performance Measures

Since no single indicator can ever be expected to signal all the trends, it is essential to combine a set of indicators so that an overall picture of the market can be built up. All the indicators used here are functions of the time and price history. These indicators can be separated into different classes depending on their usage. The main groups are

- the **trend following** indicators, which allow the trader to stay in the major directional moves (for example buy when the market is growing),
- the **overbought/oversold** indicators, which allow the trader to take contrarian positions, that is a position inverse to the current trend motivated by the fact that the market has evolved too far in a given direction (for example, sell before the market goes down),
- the **market mood** sensors (which specify to stay out of the market during specific periods).

An indicator is generally compared with a corresponding threshold value, or "break-level", usually a constant value. With some self-adjusting behaviors, the break-level can be another indicator.

The construction of these indicators is generally based on the concept of *momentum*. A momentum of the price $x$ is computed according to

$$I_{x,\tau} = x(t) - \mathrm{EMA}_x(\tau, t) \qquad (8.1)$$

The first term $x(t)$ is the price at time $t$ and the second term is some form of moving average of past prices computed during a time interval $\tau$.

A momentum is a quantity which is assumed to indicate the current market trend. It is positive when the direction of the market is up and negative when it is down.

The major problem in optimizing trading models is to avoid *overfitting* caused by the presence of noise (see also Chapter 2, Section 2.5.3). Overfitting means building trading models that fit a price history (the *in-sample* data) so well that it is no longer of general use: instead of modeling the principles underlying the price movement, it models the specific price moves observed during a particular time period. Such a model usually exhibits a totally different behavior or may fail to perform for the data not used during the optimization process (the *out-of-sample* data). To avoid overfitting during optimization, we need a good measure of the trading model performance, and robust optimization and testing procedures. Here we briefly indicate the main idea (cf. [46] and references therein for details).

The *total return*, $R_T$, is a measure of the overall success of a trading model over a period $T$ (i.e. the profit earned by the trader). It is defined by

$$R_T \equiv \sum_{j=1}^{n} r_j \qquad (8.2)$$

where $n$ is the total number of transactions during the period $T$ and $j$ labels the transactions. The quantity $r_j$ is the return associated with the $j$th transaction defined as

$$r_j = (P_{t_j} - P_{t_{j-1}})/P_{t_{j-1}}$$

where $P_{t_j}$ and $P_{t_{j-1}}$ are the current and previous transaction prices.

The total return could be a possible fitness function for the problem. However, in financial applications, it is important to know whether a given strategy has some risks associated with it or not. A high risk penalizes a high profit model and reduces its fitness. Intuitively, the risk expresses the uncertainty to realize the expected profit. It is a measure of the fluctuations of profit of the strategy during the year. A risk-sensitive fitness function which accounts for the previous criteria can be defined: it is called $X_{eff}$, the effective fitness measure of a trading model [46].

### 8.4.2 Parallel Evolution of Trading Models

For trading strategies in the GP framework, the terminal set could be financial indicators or numeric constants associated with these indicators and the function set could serve to encode the different trading rules.

Though GP trees can, in general, be constructed out of any combination of terminals and functions, imposing certain restrictions reflecting the symmetry of the problem can greatly help in the formation of meaningful solutions. As mentioned, the trading models for foreign exchange considered here are a function of the price (exchange rate) history. The price $x$ used by the indicators presents an exact symmetry $x \to -x$ corresponding to the interchange of the expressed and exchanged currencies. It is desirable that a trading model satisfies this symmetry, too. The recommendation $g$ generated by a trading model may be considered as a function $g(x)$. Enforcing the symmetry condition on the trading model thus requires the antisymmetric recommendation signal

$$g(-x) \;=\; -\,g(-x)$$

Maintaining this property in a GP tree requires tracking the symmetry property at individual nodes in the tree, and forms the basis for defining syntactic restrictions.

As noted before, the indicators and constant values appear in the GP tree as terminals. All terminal and function nodes return a real value but to enforce the symmetry $x \to -x$, three possible types are defined for reals:

- antisymmetric type **A** (for example, a price): $A(-x) \;=\; -A(-x)$,
- symmetric type **S** (for example, a volatility): $S(x) \;=\; S(-x)$,
- constant type **C** (numeric constant).

Where "volatility" is a quantity related to the standard deviation of the price changes in a given time window.

In specifying a GP trading model, every node evaluation is considered to return both a value and a symmetry type (A, S, C). The typing mechanism and syntactic restrictions are used both to assure consistent arithmetic and to categorize the symmetry properties.

For instance the addition of a momentum (antisymmetric type) and a volatility (symmetric type) provide a noisy output signal which has no meaning and will generate poor and unstable solutions. To avoid such problems, the addition is allowed only for a certain number of combinations of arguments:

| + | A | S | C |
|---|---|---|---|
| A | A | - | - |
| S | - | S | S |
| C | - | S | - |

In this table the first column is the type of the first argument and the first line corresponds to the type of the second argument. The intersection of the other columns and lines contains the type of the output signal. Here the (-) represent the combination of arguments which are not allowed.

A variety of functions may be considered for formulating trading models. Each, however, must be specified in terms of the syntactic restrictions relating to symmetry, which guide their combination with terminals and other functions. The syntactic restrictions add a type to each branch of the tree, and the operators are defined only for certain combinations of types. In this study we have used the four arithmetic functions ($+,-,*,/$), the basic comparison ($<$, $>$), and the logical operators (AND,OR,IF). Both comparison and logical operators are defined for a ternary logic {-1,0,+1} which correspond to the signal returned by a trading model. As in the case of the arithmetic functions, these operators are used with syntactic restrictions to conserve the overall symmetry properties of the generated GP trees.

The terminals used are:

- Antisymmetric indicators: $M_n$ which represent the momentum of price $x$ over $n$ days. Here we consider three different ranges: $M_8$, $M_{16}$, $M_{32}$.
- Symmetric indicators: $V_n$ that are volatility indicators over $n$ days: $V_8$, $V_{16}$
- Constant values in the range [-2, +2].

For the purpose of reducing overfitting, each trading model is tested on many exchange rate time series. The fitness measure, $X_{eff}$ of a GP-tree is then defined as the average fitness over each exchange rate, decreased by a penalty proportional to the standard deviation of these values.

Trading model evolution takes place on five exchange rates (USD-DEM, USD-JPY, USD-CHF, GBP-USD, and USD-FRF) where each time series contains 9 years of hourly data from January, 1, 1987 to December, 31, 1995. The data is divided into alternate training (in-sample) and test (out-of-sample) periods of a-year-and-a-half each.

The distributed GP model used is the asynchronous island model on a network of workstations using Parallel Virtual Machine (PVM) or Message Passing Interface (MPI) communication library routines, with steady state reproduction (see Sections 8.3.3 and 8.3.5 in this chapter). The migration of good individuals across the different sub-populations is done asynchronously with a random selection of the destination island. Given the time-consuming nature of the calculation, the choice of a parallel or distributed architecture for this problem is nearly mandatory.

Several independent optimization runs were performed in this study. In each run, a number of subpopulations independently and simultaneously evolve and each one of them periodically sends 5% of its best individuals to another, randomly selected, sub-population. A given sub-population accepts its migrants when the difference between its best and its average local fitness is smaller than half the standard deviation of the fitness. The new individuals replace an equal number of low-fitness ones in the receiving population. The evolution is stopped when a maximum number of 10000 individuals have been evaluated. The selected solution is the best solution found among all the subpopulations.

As an example, the solution which provides the best in-sample result is given by the GP-tree written in Lisp-like form:

```
(IF (> V8 0.027)
    (IF (> V16 0.108)
        (+ M16 M32)
        (* M32 1.957)
    )
    (* M8 1.542)
)
```

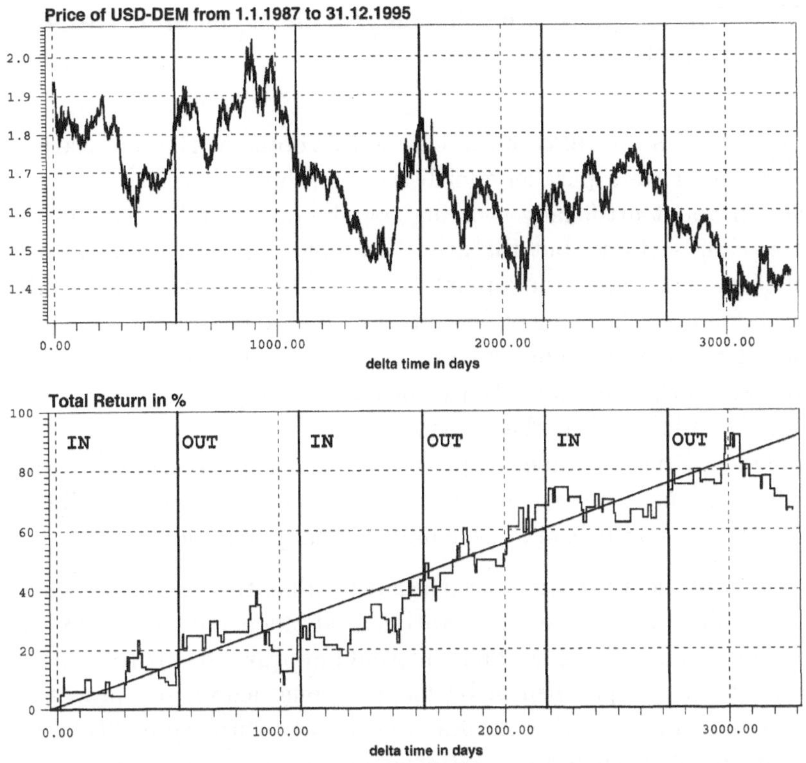


Figure 8.7 Trading results for USD-DEM exchange rate obtained with the best evolved GP-tree (see text).

Figure 8.7 presents the results obtained for the USD-DEM exchange rate with this GP-tree. The price is plotted in the top graph and the total return (in %) is given in the bottom graph. The different in-sample and out-of-sample bands are also displayed on the return graph. We can observe that the two larger periods of losses are in the out-of-sample which is a sign of overfitting.

Although the evolved solutions are relatively simple, we can conclude that reasonable trading models can readily and efficiently be evolved by distributed genetic programming without human intervention. Indeed, within their level of complexity, the evolved trading models were comparable in quality to state-of-the-art solutions but only took a fraction of the time to be determined.

## 8.5 Parallel and Distributed Artificial Neural Network Models

AT THE BEGINNING of this chapter, we observed that natural systems have a parallel and distributed dimension. We talked at length about artificially evolving populations of individuals in the preceding sections, but the prime example of massive natural parallelism is of course the animal brain. Highly parallel aspects have been found in the firing of neurons in cognitive processing and memory. These have been observed by neurobiologists in connection with any kind of perceptual activity such as vision and audition. ANNs have been inspired by their natural counterparts, therefore it is sensible to ask if the parallelism present in the brain's style of processing information can be exploited in a manner similar to what we did for the natural parallelism of evolving artificial populations.

Artificial neurons can be seen as a network of tiny processors where each unit can compute its activation level independently of the others and fire accordingly. Thus, it would appear obvious to take advantage of this parallelism by mapping one neuron per processor. However, this extremely natural view clashes with the reality of existing parallel hardware. Artificial neural networks feature many neural cells having many incoming and outgoing connections to other cells in irregular geometric patterns over a wide range. In fact, ideal architectures should be three-dimensional to allow for arbitrary interconnections. Digital hardware, on the other hand, being restricted to a two-dimensional structure, does not allow wires to pass over or touch one another. Of course two-dimensional structures can be interconnected and combined but only using long wires and slower data paths. Unless the network has a regular structure and is sparsely connected there will be a large architectural gap with ordinary parallel computer processor topology and interconnections, where regular architectures and communication patterns are the rule (see Section 8.2). This state of affairs causes mapping difficulties and it is at the origin of possible bottlenecks when trying to capture the intrinsic natural parallelism of neural nets on current parallel hardware. Indeed, the neural style of processing is a new computation paradigm that should be realized in an adequate implementation medium if it is to fulfill its potential. We will have more to say about neural style machines later. For the time being, we will concentrate on the simulation of parallel ANN models on standard parallel and distributed general-purpose computers. In fact, despite constant advances in speed of general-purpose workstations, sequential simulations of parallel ANN processes may be too slow in practice. Actual parallel implementations are thus interesting in themselves.

¿From a pragmatic point of view, it appears that there are two main levels at which artificial neural networks can be parallelized on current parallel and distributed machines: the *geometric* or architecture level and the *training* or data partitioning level. Before embarking upon the description of these possibilities, we present a simple and useful model which is similar in spirit to the master/slave architecture for evolutionary computation. It consists in farming out a number of different ANNs to different processors where the nets are trained or are working in the generalization phase. A central master process manages the slave processes and harvests the results. A simple optimization strategy can be implemented on the master such that the net with the best generalization properties is eventually retained. Of course, the optimization strategy can be based on evolutionary algorithms, in which case we fall directly into the subject matter of Chapter 4 where the topic has been discussed at length.

### 8.5.1 Geometric Parallelism

Under this heading, we try to exploit the spatial parallelism of the network by somehow dividing it into a set of subnetworks and by assigning different pieces of the network to different processors. The inherent spatial parallelism can be exploited both in the so-called "recognition" mode in which the net has learned and is employed as a pattern finder on new data sets, and in the "training" mode, where the network updates its weights and structure in order to adapt to the example data sets. Both operating modes are amenable to parallel computation but the second case is more computationally intensive and it is thus a prime candidate for parallelization.

One possibility is to use an SIMD machine and to map one neural unit per processor. This has been done for the Connection Machine [29] among others, with results that are not totally convincing. In these architectures, the problem lies with the large difference in communication speed between directly connected processors and more distant ones. Thus, although SIMD machines can be useful for regular and locally connected network architectures such as cellular neural networks, they cannot be seen as general-purpose parallel neural network simulators.

If a distributed memory MIMD machine is used, the partitioning of the net and its mapping onto the available processors can be done in a more flexible way as a function of the neural network architecture. However, for arbitrarily interconnected neural networks, the problem of optimally partitioning the neural network is analogous to a class of NP-hard graph problems and thus there are no nice algorithms for it. Fortunately, most neural networks of interest do possess some regular spatial features, such as multi-layer feedforward or Kohonen nets. We will first describe parallelizing methods for specific architectures and, since it is the most common architecture, let us start with feedforward multilayer nets. We saw in Chapter 2 that the main operation on artificial neurons is a matrix-vector multiplication. These operations appear in the update of neurons states from the weighted sum of their inputs: the vector contains the states of the units, the matrix contains the connection weights determining the reciprocal influences of the nodes, and the resulting output vector represents the activations at the output units. Thus, since matrix-vector products are well-known operations in parallel computing, the parallel partitioning schemes that have been devised for these operations in other fields can also be used in neural computation. However, in the general case of arbitrary network connections the resulting matrices are irregular and it is thus difficult to parallelize the calculation efficiently. On the other hand, either completely connected networks of the Hopfield type or regular architectures such as multi-layer feedforward give rise to dense and block matrices respectively, both of which are easily amenable to parallel computation.

### 8.5.2 Training Data Parallelism

During learning, the net is typically presented with a number of training data, either input patterns and their associated target output patterns in supervised learning, or just input data in unsupervised learning techniques. In either case, the input data sets tend to be large and the process repetitive. Thus, the neural network training phase offers scope for parallelism. For example, there have been a number of studies dealing with the parallelization of the popular backpropagation learning algorithm (Chapter 2, Section 2.5). In one approach [255] the neurons in each layer are "vertically" partitioned into different subnetworks and mapped as evenly as possible onto the available processors, as schematically depicted in Figure 8.8. A fully connected multilayered feedforward network is assumed, which means that all the activation or error values of the preceding or following layer are required in the forward and backward phases of the algorithm. This causes in turn a large amount of interprocessor communication with its associated overhead. The amount of communication can be reduced if each processor stores both the input and output weight values and computes them independently. However, the gain in communication overhead is partly offset by the redundant computations.

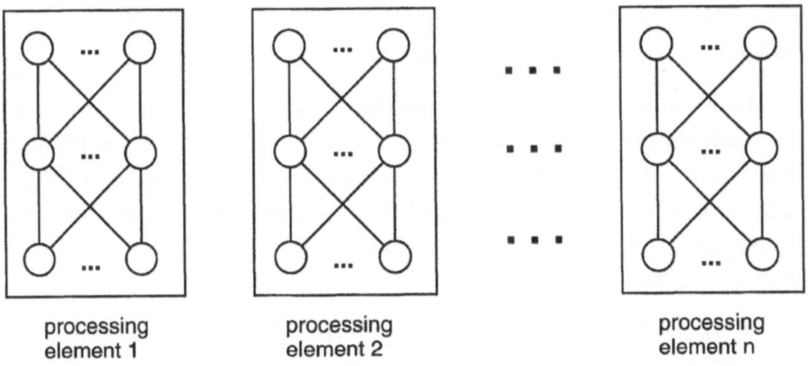


Figure 8.8 Geometric partitioning. Each processor takes care of a piece of the entire network. The network is a multi-layer, feedforward architecture where each layer is completely connected to the next one. For clarity, inter-layer connections among units that are mapped onto different processing elements are not shown.

Again using MIMD machines with distributed memory or a cluster of workstations, a simple and practical solution for speeding up the time-consuming learning phase consists in mapping $N$ copies of a neural network onto $N$ different processors (see Figure 8.9). Each processor is presented with a subset of the training examples and locally calculates its weight updates, for example using backpropagation. Subsequently, each processor sends its calculated weight changes to a special processor where all the changes are averaged out and sent back to each neural network for updating. Alternatively, an all-to-all communication between processors allows them to update their local weights by averaging all the changes. In this parallelization scheme, all the networks operate with the same weight values in order to find a solution that takes into account all the training patterns, not just a particular subset of them. A drawback of this method is that the learning algorithm is modified since the same weight matrix is used with a whole block of examples. Presenting the training data by blocks may alter the convergence properties of the backpropagation process. This effect can be countered by adapting the learning rate or by applying the weight updates immediately in a local manner and sending the global changes after a whole block of examples has been dealt with. Another problem is the overhead associated with the global weight update communication, although this can be made small if the training time is much larger than the weight transmission time. Examples of this approach to parallel neural learning can be found in references [166, 165].

When the weights are fixed and the network is used in the generalization phase or as a pattern finder on new data, the parallelism that can be achieved by farming out a different block of examples to each copy of the network is ideal since no weight changes communication/redistribution is needed and almost linear speedups can be achieved [166].

The special case of Kohonen's self-organizing feature map (see Chapter 2, Section 2.6.3) is also sufficiently typical and important to search for parallel algorithms. Remember that the self-organizing map has a vector of input neurons completely connected to a two-dimensional grid of output neurons. For map formation i.e., for unsupervised learning of the topological features of the input data, training should take place in a neighborhood centred around the most active node. The learning rule implies finding the "winning" neuron $k$, that is the output neuron with highest activation value for a given input pattern. A neighborhood $N_k$ is defined such that a weigth update is applied to the winning neuron $k$ as well as to all the units belonging to $N_k$. As training goes on the learning rate and the neighborhood size are decreased. Finding the winning unit for a given input is a compute-intensive activity and the unsupervised training itself, that is the presentation of examples, is also a slow process. Both of these phases offer scope for parallelism in order to

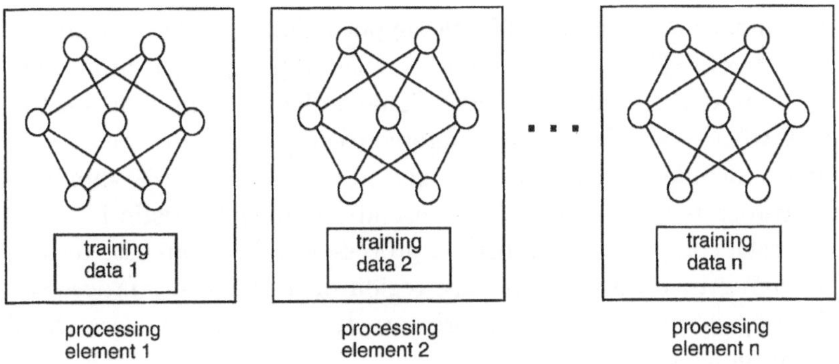


Figure 8.9 Illustration of training parallelism. The same network is mapped onto the available processors. Each network sees a portion of the training data sets. After a block of data has been processed, the processors exchange the weight updates, determine the average weight changes, and communicate them back to each other (see text).

speed up the calculations. For instance, the neurons in the input and output layers could be partitioned and mapped onto different processors as in the spatial parallelization of the back-propagation algorithm. Each processor takes care of the update of the weights in its portion of network. However, the partitioning of the network is not so simple if one wants to assure a good load balancing [175]. In fact, the activity of the different subnetworks and thus of the corresponding processors, will depend on the distribution of the traning data. The efficiency can be improved if examples are treated by blocks but, as in the case of feedforward networks trained by back-propagation, averaging the weight changes may lead to modifications in the convergence of the learning process which could even occasionally fail.

Finally, we mention parallelization methods that can be employed for more general neural network architectures. We said earlier that the problem of parallelizing network calculations for arbitrary interconnection graphs is a hard one. However, some research work has been done on heuristic rules for mapping less specific neural network classes on distributed memory multicomputers. For example, Ghosh and Hwang proposed in [74] that the neurons in a neural network be distinguished into groups, depending on their connectivity properties. Connectivity in a group of units is measured as the ratio of the number of actual connections among the units in the group to the number of connections in the hypothetical completely connected network. Thus, "core regions" are those groups of neurons that are more densely interconnected. The "influence region" is a set of cores with a sizeable number of intercore connections and "remote regions" include the other, less connected units. In order to minimize interprocessor communication a useful mapping heuristic is to assign each core region to a distinct processor and to map core regions belonging to the same influence region on neighboring processors. Clearly, modular neural network structures fit such a hierarchical partitioning scheme best.

To end this section, we would like to say something about parallel neural networks environments from the point of view of the end user. Writing parallel algorithms is a hard task in itself. The strong influence of specific architectural aspects and the fact that few standardized parallel programming languages exist adds to the difficulty of the problem. A user-friendly graphical user interface will enhance the usability of any parallel simulator both for novices and expert users. A number of commercial and public domain software packages are available. Since this is a very specific and platform-dependent topic, the reader is referred to the relevant literature, for example reference [147].

## 8.6 Parallel Neurocomputers

FROM the preceding discussion it appears that parallel implementations of neural network computing and learning algorithms present a number of advantages. Simulation of these algorithms on a general-purpose parallel or distributed architecture permits the testing of many different models at low cost, just the cost of reprogramming them, usually with the help of suitable software tools. In this way, the parallel efficiency of several different models can be analyzed without committing too early to an inflexible architecture. Another argument is that off-the-shelf processors get better and better each year and modern workstations offer good graphical user interfaces and standard tools, which is a definite advantage in terms of ease of use. Thus, when developing new ANN models or testing them on a new application, general-purpose machines offer more flexibility and are preferable. On the other hand, we already noted above that current parallel and distributed hardware is in general a poor match for the intrinsic features of neural computation. The main issue is connectivity. ANNs and their biological counterparts have many, possibly irregular connections to more or less distant units but these irregular structures are difficult to map on parallel machines and, when they are, the resulting computation is inefficient. Thus, since even parallel simulation is often too slow, it makes sense to build specialized hardware for neural computation. Even so, with current VLSI design methodologies designers are limited to a few levels of wire interconnect, each restricted to two dimensional layers. As Bailey and Hammerstrom [16] have shown, modeling a million unit ANN circuit with a thousand connections per unit, which are modest figures by biological standards, would require an inordinate amount of silicon area to be implemented directly under these technological constraints. A calculation showed [16] that a multiplexed interconnection structure, whereby several connections share the same wire, would reduce the amount of silicon by a factor of a hundred without affecting performance too much. In fact, very long wires are avoided and most networks are indeed only sparsely activated. Thus, true three-dimensional, densely connected artificial neural architectures matching the structure of biological neural nets will have to wait until new, more suitable technologies become available.

Describing detailed hardware implementations and all the related technological issues would require almost another book and we do not have space for this. Nevertheless, in the following we give a brief review of the main ideas and architectural paths that have been followed. For a more complete description the reader is referred to the book [214] which also covers parallel models for general-purpose computer architectures.

Neurocomputers have been implemented using several techniques: digital and analogue electronics, optical, and biochemical. Of these, only designs based on electronics and, to a lesser extent, on optics, are feasible today for commercial products.

### 8.6.1 Digital Neurocomputers

Digital designs are essentially of two main types: those that use commercial processors and those that employ customized hardware. The first solution is easier and faster to build and offers more flexibility in terms of programmability and software environment. Custom designs permit significant improvements in performance at a lower cost since some of the hardware that one finds in standard processors can be dispensed with for ANN computation. On the other hand, custom architectures are inflexible and cannot usually run new algorithms or more efficient versions of old algorithms without massive restructuring and redesign. Therefore, custom ANN machines usually have a short time span and are more suitable for routine applications that need high performance but in which no change is expected. Another possibility is to use *reconfigurable* hardware devices such as Field-Programmable Gate Arrays (FPGA) (see Chapter 4, Section 4.6). FPGAs can be quickly configured by the user so as to implement a given function. This capability is useful for prototyping and for testing different solutions without committing to hard-wired silicon too early. But reconfigurable devices are also interesting for the implementation of ANNs because fast circuit reconfigurability potentially brings about nearly instant hardware adaptation. This capability can be exploited for ANN hardware evolution, as we saw in Chapter 4, but other forms of adaptation are also possible. See reference [174] for further details.

### Neuro-machines Based on off-the-shelf Processors

Although standard processors have been used too, Digital Signal Processor (DSP) chips are often preferred in the design of neurocomputers. These chips usually have fast matrix-vector and matrix-matrix multiply circuits, programming support, and other features that make them suitable for ANN computation. One representative architecture of this class is the Ring Array Processor (RAP) machine [150]. The RAP is a high-performance multi-processor neurocomputer built using DSP chips. Each processor has its own local memory and a low-latency ring interconnection network, as schematically depicted in Figure 8.10. The ring connection topology was preferred to the traditional broadcast topology with a single shared bus for performance reasons, the latter suffering from well-known overheads due to bus arbitration and sharing. Multilayer feedforward ANN architectures can be easily mapped to the RAP architecture by placing one or more nodes on each processor. The most significant RAP application has been in high-speed continuous speech recognition.

A modern design using a standard RISC processor augmented with a fixed-point vector coprocessor is Berkeley's CNS-1 [9]. The machine, still under development, is targeted for for speech and vision

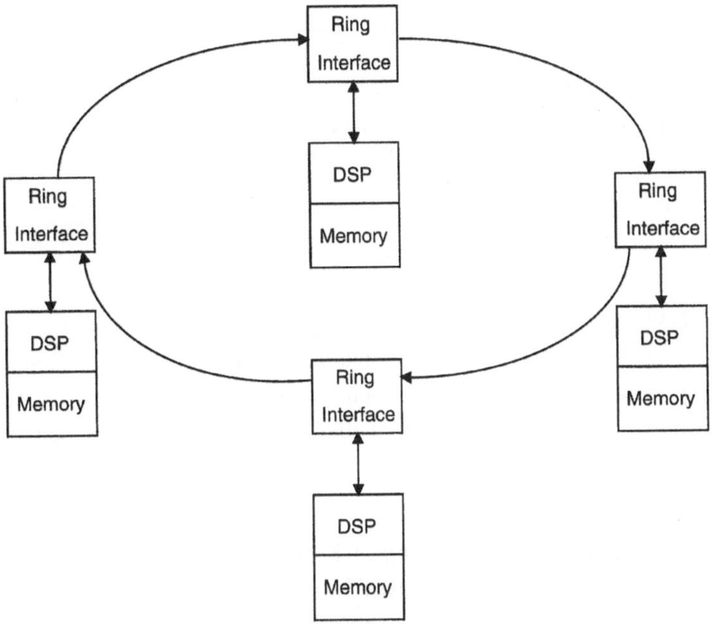


Figure 8.10 The Ring Array Processor (RAP) architecture. Each node consists of a Digital Signal Processor (DSP), local memory, and a ring interface. Messages travel in only one direction at high speed. The actual machine can have up to 64 nodes.

processing. The processing nodes (up to 1024) are connected in a mesh topology and operate in MIMD mode. High-bandwidth interprocessor processor communication and I/O will be provided to match the speed of the calculating nodes. The environment includes commercial and custom software tools for program development and testing. The machine should be one of the more powerful neurocomputers in existence.

### Neuro-machines Based on Custom Processors

In a standard microprocessor or a DSP a large amount of resources are devoted to general-purpose calculations. However, it is recognized that fixed-point arithmetic and lower precision are often sufficient for neural computation. As well, MIMD-style calculation is not always required and SIMD operation may be sufficient. The latter simplifies the design notably since a processor does not need to have its own program storage and sequencing hardware, only one instruction storage unit and one sequencing unit being needed for all the processors (see Section 8.2).

An example of a custom neurocomputer built along the previous lines is the CNAPS architecture [84] (see Figure 8.11). CNAPS features 64 processing elements per chip connected in a one-dimensional structure and functioning in SIMD mode. Each processor has a fixed-

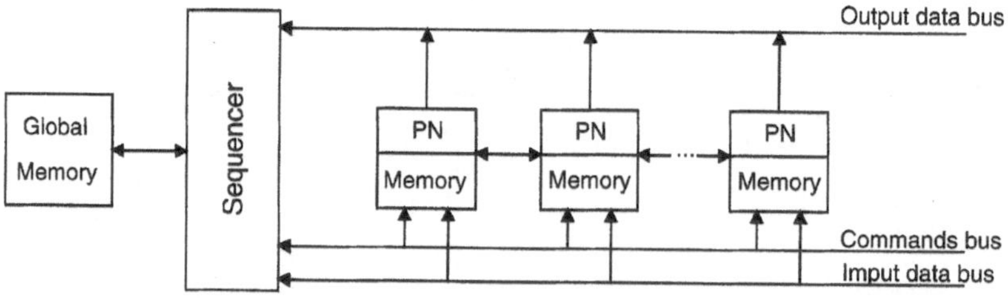


Figure 8.11 CNAPS architecture. An array of dedicated processors scaleable from 16 to 512 nodes. Each node has arithmetic circuitry, a small local memory, connection logic to the neighbour nodes and interfaces to the global busses. The machine functions by broadcasting the same instruction to all nodes in lockstep SIMD mode.

point arithmetic unit and a small local memory in which synaptic weights and other values are stored. There is a single broadcast bus for data transfer allowing many-to-many and many-to-one communication. Multiple chips can be combined to build larger systems and a CNAPS card can have up to 512 processors. The software includes several pre-programmed applications of neural networks and signal and image processing. Software tools are available for developing new algorithms through a specialized library of hand-coded standard functions.

### 8.6.2 Analog Neurocomputers

In analog computation [140, 229] data is represented and manipulated as voltage or current values as opposed to the more familiar digital representation in which signals are transmitted as streams of 0s and 1s. Analog technology has a number of advantages but also some drawbacks. Analog circuitry has greater computational density that is, operations can be done with a much lower number of transistors, thus using a smaller silicon area and power consumption is also lower. On the other hand, analog circuits are more difficult to design and are more sensitive to noise and faults than digital ones. Another limitation of analog VLSI is that algorithms are generally hard-wired and thus much less flexible than digital designs. In neural computation analog circuits seem to be most useful for building neuromorphic systems such as artificial retinas for vision that do preliminary data processing and feature extraction. According to their proponents, the relatively unprecise and massively parallel analog neurons offer a way for exploring a more biological-like artificial computation since, notoriously, biological neurons have to cope with limited precision and reliability. In particular, photoreceptors such as artificial retinas [131] and auditory receptors (artificial cochleas) [140] have been built using analog VLSI technology. The current view is that neurocomputing systems of the future will probably use analog neuromorphic components directly interacting with the environment for low-level "perceptive" processing and future extraction. Conversion to digital form will allow for further, more elaborated and precise pattern matching and classification.

### 8.6.3 Optical Neurocomputers

Optical technology, which is based on the representation and transfer of information by using photons i.e., light instead of electrons, is a very promising technology. Optics is already used with advantage in the data transmission domain where optical fiber is displacing copper wire in many applications due to its superior bandwidth and reliability and inferior weight and cost.

Light offers essentially instantaneous communication over short distances, noninterference among intersecting optical paths, and insensitivity to electromagnetic noise since photons have no electric charge. Moreover, since photons do not affect each other, highly parallel optical flows can be used to implement naturally parallel primitive computing operations in three dimensions. Optical technology has been demonstrated in the laboratory for ordinary computing, although it has not yet made a major impact in the commercial area.

For neural machines, optics is even more attractive since it promises to relieve the connectivity problem which has been repeatedly mentioned as one of the most important limiting factors in VLSI implementations of neurocomputers. Indeed, using optical systems one would be able to implement much larger-scale neural networks, to the order of $10^5$ units and $10^{10}$ interconnections. Current experimental devices are not purely optical but rather *optoelectronic* since they usually combine optical and electronic components in the same system. This means that devices that are able to convert electrical signals into optical signals and vice versa are needed and this is a new technology which is difficult to master. Most optical implementations of ANNs are of the analogous type and use optical intensity values to represent signals. The key interconnection element between neuron arrays communicating through light is a medium that can store volume holograms. Weights can be updated simultaneously during learning and fully parallel matrix-vector multiplication is possible.

Why then, if optical technology is so convenient, are there no optical computers available on the market today? The reasons are many but essentially they have to do with the difficulty of interfacing optics with electronics and with manufacturability of components, including packaging and miniaturization techniques. All these difficulties have been solved long ago by the electronics industry which explains the low production cost and popularity of VLSI chips. Nevertheless, although many issues have yet to be explored in the optical implementation of neural networks, the field has the potential for producing highly parallel, large-scale efficient neural machines. The reader can explore further this highly technical field by referring to a pair of review articles by Tanguay and Jenkins [217, 218].

Chapter 9

# Epilogue

## 9.1 Where We Stand

THE MAIN theme of this book has been the integration and the synergistic cooperation of a few important soft computing methodologies that have their roots mainly in natural systems: artificial evolution, artificial neural networks, and fuzzy systems.

While each one of these techniques can be successfully employed in isolation, we believe that in the previous chapters we have made a convincing case for the usefulness of the combined approach. On the other hand, it is by no means implied that what we have presented here is the only possibility or even the most sensible one — several other approaches are possible. We have purposefully limited ourselves to only a portion of the spectrum of the available soft computing methodologies and, obviously, to an even more restricted set of their possible combinations. As stated in the preface, our guiding principle has been the fact that the methodologies we have chosen combine naturally and that a rich corpus of successful applications exists already, some of which have been described here.

To summarize: artificial neural networks offer the possibility of inferring unknown input-output relationships by supervised and unsupervised learning from existing data. As such, they are excellent at pattern recognition and function approximation. Artificial evolution works on populations of possible designs by informed trial and error, using selection of the most adapted to bias the random process towards good solutions and building new solutions by way of pseudo-genetic operators such as crossover and mutation. This helps in solving problems that are too hard for standard deterministic methods provided that we are content with satisfactory rather than optimal solutions. Finally, fuzzy systems imitate human problem-solving strategies by using vague and imprecise language constructs in a way that can be expressed numerically and is thus open to standard computation.

Each methodology has its strong and weak points and each can be augmented and strengthened when it is suitably combined with one or more of the others. For example, connection weights and net architectures are difficult to design for neural networks. But artificial evolution comes to the rescue by offering a number of ways in which this can be done automatically. ANNs also suffer from a kind of opacity: the net operation is difficult to understand and analyse in logical terms. By using fuzzy concepts in neural networks, their interpretation becomes easier. On the other hand, neural networks offer a satisfactory automated solution to the problem of membership function determination and fuzzy rule induction. Artificial evolution has also been frequently used to automate the process of finding good fuzzy rule-based systems. Reciprocally, fuzzy systems have been used to automatically tune parameters and monitor the evolution. As well, since sometimes the fitness function on which individual selection is performed is known only approximately, fuzzy concepts may help in defining it.

Natural problem solving strategies are pragmatic and opportunistic and, when consistently successful, they become part of the repertoire of the individual and of the population by way of evolution, learning, and cultural transmission. This brings us to another point in system design and problem solving that we have implicitly stressed in the book and which we believe is of general value. Indeed, we think that it is pointless to argue over the superiority of one methodology over another. In the past we have witnessed a number of religious wars among scholars such as connectionist versus symbolic AI. In our opinion, this is a waste of time and energy since, in general, each problem solving approach has its own niche in which it is more effective and we should choose the one that is most suitable for a given problem. When no single method is the best for a given problem spectrum, then the most sensible thing to do is to suitably combine two or more methods each offering good performance in a given area or for a given purpose in the overall solution process.

In fact, most real-life problems have one common feature: they are complex and full of interrelations and constraints between their constituent parts. We believe that in many cases no single technique is enough by itself for an efficient and satisfactory solution. Complex system analysis can only be achieved by decoupling and hierarchically structuring the solution methodology. Cooperative and hybrid soft computing techniques may certainly help in this respect.

By the way, in this book we have concentrated on cooperative approaches in which one soft computing methodology is used to enhance or to counter a weakness of another. However, we have also said that several techniques can be used sequentially to solve parts of the problem in a "pipeline" fashion, so to speak. This is already common practice in engineering and application fields and will increasingly be used.

## 9.2 Future Prospects

TO BE SURE, we do not have a crystal ball with which to look into the future, but we think that it is fair to say that soft computing methodologies will play an even more important role. There are at least three main reasons to anticipate that:

- Hard computing often fails to solve problems because of an unnecessary emphasis on precision. If we relax precision, we can achieve tractability and lower solution cost;
- Computing power is becoming cheaper and cheaper, and there is no reason to expect that this trend will not continue: as more computing power is available, soft computing, which is in general more computationally expensive than conventional algorithms, becomes more efficient and attractive, for all users.
- Ecology and economy (meaning sustainable development and effective use of limited resources) will increase their importance in the years to come, and any techniques capable of helping us to better organize, save, and dispose of resources, like soft computing, will gain the center of the stage.

Indeed, as the years pass, our ambitions expand, and we want to tackle and solve problems of ever increasing complexity and difficulty. As a consequence, we will witness a growing demand for design, control, and problem-solving techniques capable of dealing with incomplete information, noise, and possibly time-varying aspects, all of which are present in complex production, massive data analysis, and ecological or economical systems. The solutions will also have to be adaptive and resistant to failures. It appears that these features are easier to take into account in the framework of soft computing, especially in cooperative and hybrid approaches. There are already clear signs that soft computing techniques are being increasingly used on an industrial scale to solve difficult application problems. This shows the maturity of the field and we believe that the trend will intensify.

Soft computing-based hardware devices of low to medium complexity such as neural networks and fuzzy controllers have been on the market for some time, especially in Japan, without much fuss since this hardware has been quietly included in many consumer electronic goods. We foresee that this trend will continue and that more and more functions in production plants and in technological products will be taken up by soft computing software and hardware. This includes predictive and anticipatory control systems as well as more flexible and better computer-human interfaces that take into account the imprecision and vagueness of human language and adapt to user preferences and requirements.

Many of these developments will tend to be idiosyncratic and special-purpose. Far from being a drawback, this will allow the customization of better designed devices. Of course, general-purpose computing devices will always have their place, but more specialized machines will become more common.

In a more speculative vein, foreseeable technological advances might make this progress even faster and more spectacular. If ways are found to build and design high-speed, dynamically reconfigurable circuits that are faster and more flexible than today's FPGAs, and especially if we learn to master the technologies required to build molecular and nanotechnological-based devices, then the way will be open for very effective, microscopic, extremely cheap pervasive machines that could be built by means of a combination of design and soft computing techniques in a sort of evolutionary self-organizing engineering. These machines would incorporate soft computing algorithms, including fault-tolerance, adaptive, and reconfiguring capabilities directly in hardware and open up completely new horizons and prospects.

# Bibliography


[1] E. H. L. Aarts, A. E. Eiben, and K. M. van Hee. *A General Theory of Genetic Algorithms.* Computing Science Notes. Eindhoven University of Technology, Eindhoven, 1989.

[2] D. H. Ackley and M. L. Littman. Interactions between learning and evolution. In C.G. Langton, J.D. Farmer, S. Rasmussen, and C. Taylor, editors, *Artificial Life II: Proceedings Volume of Santa Fe Conference*, volume X, pages 487–509. Addison Wesley: Series of the Santa Fe Institute Studies in the Sciences of Complexities, Redwood City, CA, 1992.

[3] D. Andre and J. R. Koza. Parallel genetic programming: A scalable implementation using the transputer network architecture. In P. Angeline and K. Kinnear, editors, *Advances in Genetic Programming 2*, pages 317–337. The MIT Press, Cambridge, MA, 1996.

[4] P. Angeline, G. Saunders, and J. Pollack. Complete induction of recurrent neural networks. In A. V. Seibald and L. J. Fogel, editors, *Proceedings of the Third Conference on Evolutionary Programming*, pages 1–8. World Scientific, Singapore, 1994.

[5] P.J. Angeline and K.E. Kinnear Jr. (Eds.). *Advances in Genetic Programming 2*. The MIT Press, Cambridge, Massachusetts, 1996.

[6] T. Ankenbrand and M. Tomassini. Multivariate time series modelling of financial markets with artificial neural networks. In D. W. Pearson, N. C. Steele, and F. Albrecht, editors, *Proceedings of International Conference on Artificial Neural Networks and Genetic Algorithms (ICANNGA95)*, pages 257–260. Springer-Verlag KG, Vienna, 1995.

[7] T. Ankenbrand and M. Tomassini. Predicting multivariate financial time series using neural networks: the swiss bond case. In *Proceedings of the IEEE/IAFE Conference on Computational Intelligence for Financial Engineering*, pages 27–33. IEEE, New York, 1996.

[8] S. Arnone, M. Dell'Orto, and A. Tettamanzi. Toward a fuzzy government of genetic populations. In *Proceedings of the 6th IEEE Conference on Tools with Artificial Intelligence (TAI'94)*, pages 585–591. IEEE Computer Society Press, Los Alamitos, CA, 1994.

[9] K. Asanović, J. Beck, J. Feldman, N. Morgan, and J. Wawrzynek. A supercomputer for neural computation. In *Proceedings of the IEEE International Conference on Neural Networks*, volume 1, pages 5–9. IEEE Press, Piscataway, NJ, 1994.

[10] T. Ash and G. Cottrell. Topology-modifying neural network algorithms. In Michael A. Arbib, editor, *Handbook of Brain Theory and Neural Networks*, pages 990–993. MIT Press, 1995.

[11] Various authors. Special issue on evolvable hardware. *Communications of the ACM*, 42:46–79, 1999.

[12] T. Bäck. *Evolutionary algorithms in theory and practice*. Oxford University Press, Oxford, 1996.

[13] T. Bäck. The interaction of mutation rate, selection, and self-adaptation within genetic algorithms. In R. Männer and B. Manderick, editors, *Parallel Problem Solving from Nature 2*, pages 85–94. Elsevier, Amsterdam, 1992.

[14] T. Bäck and F. Hoffmeister. Extended selection mechanisms in genetic algorithms. In R. K. Belew and L. B. Booker, editors, *Proceedings of the Fourth International Conference on Genetic Algorithms*, pages 92–99. Morgan Kaufmann, San Mateo, CA, 1991.

[15] S. Baglioni, C. da Costa Pereira, D. Sorbello, and A. Tettamanzi. An evolutionary approach to multiperiod asset allocation. In R. Poli, W. Banzhaf, W. Langdon, J. Miller, P. Nordin, and T. Fogarty, editors, *Genetic Programming. European Conference, EuroGP 2000*, pages 225–236. Springer-Verlag, Berlin, 2000.

[16] J. Bailey and D. Hammerstrom. Why VLSI implementations of associative VLCNs require connection multiplexing. In *Proceedings of the 1988 International Conference on Neural Networks*, pages 173–188. IEEE, Piscataway, NJ, 1988.

[17] J. Baker. Adaptive selection methods for genetic algorithms. In J. J. Grefenstette, editor, *Proceedings of the First International Conference on Genetic Algorithms*, pages 101–111. Lawrence Erlbaum Associates, Hillsdale, NJ, 1985.

[18] W. Banzhaf, P. Nordin, R. E. Keller, and F. D. Francone. *Genetic programming, An Introduction.* Morgan Kaufmann, San Francisco, CA, 1998.

[19] A. G. Barto. Reinforcement learning. In Michael A. Arbib, editor, *Handbook of Brain Theory and Neural Networks*, pages 804–809. MIT Press, 1995.

[20] R. Belew, J. McInerney, and N. N. Schraudolph. Evolving networks: Using the genetic algorithm with connectionist learning. Technical Report CS90-174, University of California, San Diego, 1990.

[21] R. K Belew and M. Mitchell, editors. *Adaptive Individuals in Evolving Populations. Models and Algorithms.* Addison-Wesley, Redwood City, CA, 1996.

[22] K. P. Bennett and O. L. Mangasarian. Neural network training via linear programming. In P. M. Pardalos, editor, *Advances in Optimization and Parallel Computing*, pages 56–57. Elsevier Science, Amsterdam, 1992.

[23] M. Beretta, G. Degli Antoni, C. Mastrorilli, and A. Tettamanzi. Fuzzy heuristic signal compression based on linguistic variables. In F. Masulli A. Bonarini, D. Mancini and A. Petrosino, editors, *Proceedings of the 2nd Italian Workshop on Fuzzy Logic (WILF '97)*, pages 276–283. World Scientific, Singapore, 1998.

[24] M. Beretta, G. Degli Antoni, and A. Tettamanzi. Evolutionary synthesis of a fuzzy image compression algorithm. In *Proceedings of the Fourth European Congress on Intelligent Techniques and Soft Computing (EUFIT '96)*, volume 1, pages 466–470. Aachen, September 2–5 1996.

[25] M. Beretta and A. Tettamanzi. An evolutionary approach to fuzzy image compression. In F. Masulli A. Bonarini, D. Mancini and A. Petrosino, editors, *New Trends in Fuzzy Logic, Proceedings of the Italian Workshop on Fuzzy Logic (WILF '95)*, pages 49–57. World Scientific, Singapore, 1996.

[26] A. Bergmann, W. Burgard, and A. Hemker. Adjusting parameters of genetic algorithms by fuzzy control rules. In K.-H. Becks and D. Perret-Gallix, editors, *New Computing Techniques in Physics Research III*. World Scientific, Singapore, 1993.

[27] H. Bersini, J.-P. Nordvik, and A. Bonarini. A simple direct adaptive fuzzy controller derived from its neural equivalent. In *Proceedings of the IEEE International Conference on Fuzzy Systems 1993*, pages 345–350. San Francisco, CA, 1993.

[28] J. Bezdek. *Pattern Recognition with Fuzzy Objective Function Algorithms*. Plenum, New York, 1981.

[29] G. Blelloch and C. R. Rosenberg. Network learning on the connection machine. In *Proceedings of the Tenth International Joint Conference on Artificial Intelligence*, pages 323–326. Morgan Kaufmann, Los Altos, CA, 1987.

[30] A. Bonarini and F. Ludovico. A neuro-fuzzy approach to autonomous navigation for a mobile robot. In *Proceedings of the Third International Symposium on Soft Computing (SOCO'99)*. Genova, Italy, June 1–4 1999.

[31] H. Braun. Evolving neural networks for application oriented problems. In D. B. Fogel and W. Atmar, editors, *Proceedings*

*of the Second Conference on Evolutionary Programming*, pages 62–71. Evolutionary Programming Society, La Jolla, California, 1993.

[32] F. Z. Brill, D. E. Brown, and W. N. Martin. Fast genetic selection of features for neural network classifiers. *IEEE Transactions on Neural Networks*, 3(2):324–328, 1992.

[33] A. Brindle. Genetic agorithms for function optimization. Technical Report TR81-2, Department of Computer Science, University of Alberta, Edmonton, 1981.

[34] C. Caldwell and V. S. Johnston. Tracking a criminal suspect through "face space" with a genetic algorithm. In R. K. Belew and L. B. Booker, editors, *Proceedings of the Fourth International Conference on Genetic Algorithms*, pages 416–421. Morgan Kaufmann, San Mateo, CA, 1991.

[35] E. Cantú-Paz. A summary of research on parallel genetic algorithms. Technical Report 95007, Illinois Genetic Algorithms Laboratory, University of Illinois at Urbana-Champaign, Urbana, IL, July 1995.

[36] E. Cantú-Paz. Designing efficient master-slave parallel genetic algorithms. Technical Report 95004, Illinois Genetic Algorithms Laboratory, University of Illinois at Urbana-Champaign, Urbana, IL, 1997.

[37] E. Cantú-Paz and D. E. Goldberg. Modeling idealized bounding cases of parallel genetic algorithms. In J. R. Koza, K. Deb, M. Dorigo, D. B. Fogel, M. Garzon, H. Iba, and R. L. Riolo, editors, *Genetic Programming 1997: Proceedings of the Second Annual Conference*, pages 456–462. Morgan Kaufmann, San Francisco, CA, 1997.

[38] M. Capcarrère, A. Tettamanzi, M. Tomassini, and M. Sipper. A statistical study of a class of cellular evolutionary algorithms. *Evolutionary Computation*, 7(3):255–274, 1999.

[39] C. Carlsson and R. Fullér. A neuro-fuzzy system for portfolio evaluation. In R. Trappl, editor, *Cybernetics and Systems '96, Proceedings of the Thirteenth European Meeting on Cybernetics and Systems Research*, pages 296–299. Austrian Society for Cybernetic Studies, Vienna, 1996.

[40] G. A. Carpenter, S. Grossberg, N. Markuzon, J. H. Reynolds, and D. B. Rosen. Fuzzy ARTMAP: A neural network architecture for incremental supervised learning of analog multidimensional maps. *IEEE Transactions on Neural Networks*, 3:698–712, 1992.

[41] G. A. Carpenter, S. Grossberg, and D. B. Rosen. Fuzzy ART: Fast stable learning and categorization of analog patterns by an adaptive resonance system. *Neural Networks*, 4(6):759–771, 1991.

[42] D. J. Chalmers. The evolution of learning: An experiment in genetic connectionism. In D. S. Touretzky, J. L. Elman, and G. E. Hinton, editors, *Connectionist Models: Proceedings of the 1990 Summer School*, pages 81–90. Morgan Kaufmann, San Mateo, California, 1990.

[43] E. J. Chang and R. P. Lippmann. Using genetic algorithms to improve pattern classification performance. In R. P. Lippmann, J. E. Moody, and D. S. Touretsky, editors, *Advances in neural information processing 3*, pages 797–803. Morgan Kaufmann, San Mateo, California, 1991.

[44] B. Cheng and D. M. Titterington. Neural networks: a review from a statistical perspective. *Statistical Science*, 9:2–54, 1994.

[45] S. Cho. Neural-network classifiers for recognizing totally unconstrained handwritten numerals. *IEEE Transactions on Neural Networks*, 8:43–53, 1997.

[46] B. Chopard, O. Pictet, and M. Tomassini. Parallel and distributed evolutionary computation for financial applications. *Parallel Algorithms and Applications*, 15:15–36, 2000.

[47] J. P. Cohoon, S. U. Hegde, W. N. Martin, and D. Richards. Punctuated equilibria: A parallel genetic algorithm. In J. J. Grefenstette, editor, *Proceedings of the Second International Conference on Genetic Algorithms*, pages 148–154. Lawrence Erlbaum Associates, 1987.

[48] J. D. Cowan. Fault tolerance. In Michael A. Arbib, editor, *Handbook of Brain Theory and Neural Networks*, pages 390–395. MIT Press, 1995.

[49] Y. Le Cun, B. Boser, J. Denker, D. Henderson, R. Howard, W. Hubbard, and L. Jackel. Handwritten digit recognition with a backpropagation network. In D. S. Touretzky, editor, *Advances in Neural Information Processing Systems 2*, pages 396–404. Morgan Kaufmann, San Mateo, CA, 1990.

[50] C. Darwin. *The Origin of Species.* John Murray, London, 1859.

[51] T. E. Davis and J. C. Principe. A simulated annealing like convergence theory for the simple genetic algorithm. In R. K. Belew and L. B. Booker, editors, *Proceedings of the Fourth International Conference on Genetic Algorithms*, pages 174–181. Morgan Kaufmann, San Mateo, CA, 1991.

[52] R. Dawkins. *The Blind Watchmaker.* W.W. Norton and Company, New York, 1986.

[53] H. de Garis, M. Korkin, F. Gers, E. Nawa, and M. Hough. Building an artificial brain using an FPGA based CAM-Brain machine. *Applied Mathematics and Computation Journal*, 111:163–192, 2000. Special issue on Artificial Life and Robotics, Artificial Brain, Brain Computing and Brainware.

[54] D. Driankov, H. Hellendoorn, and M. Reinfrank. *An Introduction to Fuzzy Control.* Springer-Verlag, New York, 1993.

[55] D. Dubois and H. Prade. Operations on fuzzy numbers. *International Journal of System Science*, 9:613–626, 1978.

[56] D. Dubois and H. Prade. *Fuzzy Sets and Systems, Theory and Applications.* Academic, New York, 1980.

[57] D. Dubois and H. Prade. Soft computing, fuzzy logic and artificial intelligence. *Soft Computing*, 2(1):7–11, 1998.

[58] A.-P. Refenes (Ed.). *Neural Networks in the Capital Markets.* John Wiley, Chichester, 1995.

[59] K.E. Kinnear Jr. (Ed.). *Advances in Genetic Programming.* The MIT Press, Cambridge, Massachusetts, 1994.

[60] S.E. Fahlman and C. Lebiere. The Cascade-Correlation Learning Architecture. In Touretzky (Ed.) *Advances in Neural Information Processing Systems 2*, pages 524–532, Morgan-Kaufmann, San Mateo, CA, 1990.

[61] S. Fatikov and G. Wohlke. Neuro-fuzzy control approach for intelligent microrobots. In *Proceedings of the IEEE International Conference on Systems, Man and Cybernetics*, volume 4, pages 441–446. La Touquet, France, October 17–20, 1993.

[62] F. Fernández, J. M. Sánchez, M. Tomassini, and J. A. Gómez. A parallel genetic programming tool based on PVM. In J. Dongarra, E. Luque, and Tomás Margalef, editors, *Recent Advances in Parallel Virtual Machine and Message Passing Interface*, volume 1697 of *Lecture Notes in Computer Science*, pages 241–248. Springer-Verlag, Heidelberg, 1999.

[63] F. Fernández, M. Tomassini, W. F. Punch III, and J. M. Sánchez. Experimental study of multipopulation parallel genetic programming. In Riccardo Poli, Wolfgang Banzhaf, William B. Langdon, Julian F. Miller, Peter Nordin, and Terence C. Fogarty, editors, *Genetic Programming, Proceedings of EuroGP'2000*, volume 1802 of *LNCS*, pages 283–293. Springer-Verlag, Heidelberg, 2000.

[64] D. Floreano and F. Mondada. Evolution of plastic neurocontrollers for situated agents. In P. Maes, M. Mataric, J-A. Meyer, J. Pollack, H. Roitblat, and S. Wilson, editors, *From Animals to Animats IV: Proceedings of the Fourth International Conference on Simulation of Adaptive Behavior*, pages 402–410. MIT Press-Bradford Books, Cambridge, MA, 1996.

[65] D. Floreano and F. Mondada. Evolutionary neurocontrollers for autonomous mobile robots. *Neural Networks*, 11:1461–1478, 1998.

[66] D. Floreano and J. Urzelai. Evolution and learning in autonomous robotic agents. In D. Mange and M. Tomassini, editors, *Bio-Inspired Computing Machines: Towards Novel Computational Architectures*, pages 317–346. Presses Polytechniques et Universitaires Romandes, Lausanne, 1998.

[67] D. Floreano and J. Urzelai. Evolution of Neural Controllers with Adaptive Synapses and Compact Genetic Encoding. In D. Floreano, J.D. Nicoud, and F. Mondada, editors, *Advances In Artificial Life: Proceedings of the 5th European Conference on Artificial Life (ECAL'99)*, pages 183–194. Springer-Verlag, Berlin, 1999.

[68] D. B. Fogel. *Evolving Artificial Intelligence.* PhD thesis, University of California, San Diego, 1992.

[69] D. B. Fogel, L. J. Fogel, and V. W. Porto. Evolving neural networks. *Biological Cynernetics*, 63:487–493, 1990.

[70] L. J. Fogel. *On the Organization of Intellect.* PhD thesis, University of California, Los Angeles, 1964.

[71] L. J. Fogel, A. J. Owens, and M. J. Walsh. *Artificial Intelligence through Simulated Evolution.* John Wiley & Sons, New York, 1966.

[72] G. Folino, C. Pizzuti, and G. Spezzano. Genetic programming and simulated annealing: a hybrid method to evolve decision trees. In Riccardo Poli, Wolfgang Banzhaf, W. B. Langdon, Julian F. Miller, Peter Nordin, and Terence C. Fogarty, editors, *Genetic Programming, Proceedings of EuroGP'2000*, Lecture Notes in Computer Science, pages 294–303. Springer-Verlag, Berlin, 2000.

[73] P. D. Gader, J. M. Keller, R. Krishnapuram, J. Chiang, and M. Mohamed. Neural and fuzzy methods in handwriting recognition. *IEEE Computer*, pages 79–86, February 1997.

[74] J. Ghosh and K. Hwang. Mapping neural networks onto message-passing multicomputers. *Journal of Parallel and Distributed Computing*, 6:291–330, 1989.

[75] J. Goguen. The logic of inexact concepts. *Synthese*, 19:325–373, 1969.

[76] D. E. Goldberg. *Genetic Algorithms in Search, Optimization and Machine Learning.* Addison-Wesley, 1989.

[77] D. E. Goldberg. Messy genetic algorithms: Motivation, analysis and first results. *Complex Systems*, 3:493–530, 1989.

[78] D. E. Goldberg. Sizing populations for serial and parallel genetic algorithms. In J. J. Grefenstette, editor, *Proceedings of the International Conference on Genetic Algorithms and their Applications*, pages 70–79. Morgan Kaufmann, San Mateo, CA, 1989.

[79] M. Gordon. Probabilistic and genetic algorithms for document retrieval. *Communications of the ACM*, 31(10):1208–1218, 1988.

[80] J. J. Grefenstette. Optimization of control parameters for genetic algorithms. *IEEE Transactions on Systems, Man, and Cybernetics*, 16:122–128, 1986.

[81] F. Gruau. Genetic synthesis of boolean neural networks with a cell rewriting developmental process. In D. Whitley and J. D. Schaffer, editors, *Proceedings of the Workshop on Combinations of Genetic Algorithms and Neural Networks.* IEEE Computer Society Press, Los Alamitos, CA, 1992.

[82] K. Gurney. *An introduction to neural networks.* UCL Press, London, 1997.

[83] P. Hájek, L. Godo, and F. Esteva. Fuzzy logic and probability. In *Proceedings of the Eleventh Annual Conference on Uncertainty in Artificial Intelligence (UAI-95).* Montreal, Québec, August 18–20, 1995.

[84] D. Hammerstrom. A highly parallel digital architecture for neural network emulation. In J. Delgado-Frias and W. Moore, editors, *VLSI for Artificial Intelligence and Neural Networks.* Plenum, New York, 1991.

[85] H. V. Harlow. Asset allocation in a downside-risk framework. *Financial Analysts Journal*, pages 30–40, September/October 1991.

[86] M. H. Hassoun. *Fundamentals of artificial neural networks.* MIT Press, Cambridge, MA, 1995.

[87] D. O. Hebb. *The Organization of Behavior.* Wiley, New York, 1949.

[88] H. Hellendoorn and C. Thomas. Defuzzification in fuzzy controllers. *Intelligent and Fuzzy Systems*, 1:109–123, 1993.

[89] F. Herrera and M. Lozano. Heuristic crossover for real-coded genetic algorithms based on fuzzy connectives. In H.-M. Voigt, W. Ebeling, I. Rechenberg, and H.-P. Schwefel, editors, *Proceedings of the 4th International Conference on Parallel*

*Problem Solving from Nature*, pages 336–345. Springer-Verlag, Berlin, 1996.

[90] F. Herrera and M. Lozano. Adaptive genetic algorithms based on coevolution with fuzzy behaviors. Technical Report DECSAI-98105, University of Granada, Dept. of Computer Science and A.I., November 1997.

[91] F. Herrera and M. Lozano. Fuzzy genetic algorithms: Issues and models. Technical Report DECSAI-98116, University of Granada, Dept. of Computer Science and A.I., June 1998.

[92] F. Herrera, M. Lozano, and J. L. Verdegay. Fuzzy connectives based crossover operators to model genetic algorithms population diversity. *Fuzzy Sets and Systems*, 92(1):21–30, 1997.

[93] W. D. Hillis. Co-evolving parasites improve simulated evolution as an optimization procedure. In C. G. Langton, C. Taylor, J. D. Farmer, and S. Rasmussen, editors, *Artificial Life II*, volume X of *SFI Studies in the Sciences of Complexity*, pages 313–324. Addison-Wesley, Redwood City, CA, 1992.

[94] G. E. Hinton and S. J. Nowlan. How learning can guide evolution. *Complex Systems*, 1:495–502, 1987.

[95] H. Hirota and W. Pedrycz. AND/OR neuron in modeling fuzzy set connectives. *IEEE Transactions on Fuzzy Systems*, 2:151–161, 1994.

[96] A. Hobbs and N. G. Bourbakis. A neurofuzzy arbitrage simulator for stock investing. In *Proceedings of the IEEE/IAFE 1995 Computational Intelligence for Financial Engineering.* New York, April 9–11, 1995.

[97] J. H. Holland. *Adaptation in Natural and Artificial Systems.* The University of Michigan Press, Ann Arbor, Michigan, 1975.

[98] J. H. Holland. *Adaptation in Natural and Artificial Systems.* The MIT Press, Cambridge, Massachusetts, second edition, 1992.

[99] J. E. Hopcroft and J. D. Ullman. *Formal Languages and Their Relation to Automata.* Addison-Wesley series in Computer Science and Information Processing. Addison-Wesley, Reading, MA, 1969.

[100] P. Husbands. An ecosystem model for integrated production planning. *Journal of Computer Integrated Manufacturing*, 6:74–86, 1993.

[101] K. Hwang. *Advanced Computer Architecture*. McGraw Hill, New York, 1993.

[102] H. Ishibuchi, K. Nozaki, H. Tanaka, Y. Hosaka, and M. Matsuda. Empirical study on learning in fuzzy systems by rice taste analysis. *Fuzzy Sets and Systems*, 64(2):129–144, 1994.

[103] J.-S. Jang and C.-T. Sun. Neuro-fuzzy modeling and control. *Proceedings of the IEEE*, 83(3):378–406, 1995.

[104] J.-S. R. Jang, C.-T. Sun, and E. Mizutani. *Neuro-Fuzzy and Soft Computing*. Prentice Hall, NJ, 1997.

[105] C. Z. Janikow. Fuzzy processing in decision trees. In *Proceedings of the Sixth International Symposium on Artificial Intelligence*, pages 360–367. 1993.

[106] C. Z. Janikow. A genetic algorithm for learning fuzzy controllers. In *Proceedings of the ACM Symposium on Applied Computing*, pages 232–236. ACM Press, New York, 1994.

[107] C. Z. Janikow. A genetic algorithm for optimizing fuzzy decision trees. In L. J. Eshelman, editor, *Proceedings of the Sixth International Conference on Genetic Algorithms*, pages 421–428. Morgan Kaufmann, San Francisco, CA, 1995.

[108] K. De Jong. *An Analysis of the Behavior of a Class of Genetic Adaptive Systems*. PhD thesis, University of Michigan, 1975.

[109] C. L. Karr. Design of an adaptive fuzzy logic controller using a genetic algorithm. In R. K. Belew and L. B. Booker, editors, *Proceedings of the Fourth International Conference on Genetic Algorithms*. Morgan Kaufmann, San Mateo, CA, 1991.

[110] C. L. Karr. Genetic algorithms for fuzzy controllers. *AI Expert*, March 1991.

[111] A. Kaufmann and M. Gupta. *Introduction to Fuzzy Arithmetic: Theory and applications*. International Thomson Computer Press, London, 1991.

[112] H. Kitano. Designing neural networks by genetic algorithms using graph generation systems. *Complex Systems*, 4:461–476, 1990.

[113] G. Klir and T. Folger. *Fuzzy Sets, Uncertainty and Information.* Prentice Hall, Englewood Cliffs, N.J., 1988.

[114] T. Kohonen. *Self-Organizing Maps*, volume 30. Springer Series in Information Sciences, Springer-Verlag, Berlin, 1995.

[115] B. Kosko. *Neural Networks and Fuzzy Systems. A dynamical systems approach to machine intelligence.* Prentice Hall, Englewood Cliffs, NJ, 1992.

[116] B. Kosko. Fuzzy systems as universal approximators. In *Proc. IEEE Int. Conf. on Fuzzy Systems*, pages 1153–1162. IEEE Press, Piscataway, NJ, 1994.

[117] J. R. Koza. *Genetic Programming.* The MIT Press, Cambridge, Massachusetts, 1992.

[118] J. R. Koza. *Genetic Programming II.* The MIT Press, Cambridge, Massachusetts, 1994.

[119] D. Kraft, F. E. Petry, B. P. Buckles, and T. Sadisavan. Genetic algorithms for query optimization in information retrieval: Relevance feedback. In L. A. Zadeh E. Sanchez, T. Shibata, editor, *Genetic algorithms and fuzzy logic systems: Soft computing perspectives*, pages 155–173. World Scientific, Singapore, 1997.

[120] K. S. Krishnakumar. Micro-genetic algorithms for stationary and nonstationary function optimization. *SPIE – Intelligent Control and Adaptive Systems*, 1196:289–296, Philadelphia, PA, 1989.

[121] C. G. Langton, C. Taylor, J. D. Farmer, and S. Rasmussen, editors. *Artificial Life II*, volume X of *SFI Studies in the Sciences of Complexity.* Addison-Wesley, Redwood City, CA, 1992.

[122] M. Lee and H. Takagi. Dynamic control of genetic algorithms using fuzzy logic techniques. In S. Forrest, editor, *Proceedings of the Fifth International Conference on Genetic Algorithms*, pages 76–83. Morgan Kaufmann, San Mateo, CA, 1993.

[123] M. Lee and H. Takagi. Embedding apriori knowledge into an integrated fuzzy system design method based on genetic algorithms. In *Proceedings of the 5th IFSA World Congress (IFSA'93)*, pages Vol. II, 1293–1296. July 4-9, 1993.

[124] M. Lee and H. Takagi. Integrating design stages of fuzzy systems using genetic algorithms. In *Proceedings of the 2nd International Conference on Fuzzy Systems (FUZZ-IEEE'93)*, Vol. I, pages 612–617, 1993.

[125] M. A. Lee and H. Takagi. A framework for studying the effects of dynamic crossover, mutation and population sizing in genetic algorithms. In T. Furuhashi, editor, *Advances in Fuzzy Logic, Neural Networks and Genetic Algorithms*, pages 111–126. Springer-Verlag, Berlin, 1994.

[126] S. C. Lin, W. F. Punch, and E. D. Goodman. Coarse-grain parallel genetic algorithms: Categorization and a new approach. In *Sixth IEEE SPDP*, pages 28–37, 1994.

[127] A. Lindenmayer. Mathematical models for cellular interaction in development, Parts I and II. *Journal of Theoretical Biology*, 18:280–315, 1968.

[128] A. Loraschi and A. Tettamanzi. An evolutionary algorithm for portfolio selection within a downside risk framework. In C. Dunis, editor, *Forecasting Financial Markets*, Series in Financial Economics and Quantitative Analysis, pages 275–285. John Wiley & Sons, Chichester, 1996.

[129] A. Loraschi, A. Tettamanzi, M. Tomassini, and P. Verda. Distributed genetic algorithms with an application to portfolio selection problems. In *Proceedings of the International Conference on Artificial Neural Networks and Genetic Algorithms*, pages 384–387. Springer-Verlag, Wien, New York, 1995.

[130] N. Mackin. *The Development of an Expert System for Planning Orthodontic Treatment*. PhD thesis, University of Bristol, 1992.

[131] M. Mahowald. *An Analog VLSI System for Stereoscopic Vision*. Kluwer, Boston, 1994.

[132] E. H. Mamdani. Advances in linguistic synthesis of fuzzy controllers. *International Journal of Man Machine Studies*, 8:669–678, 1976.

[133] B. Manderick and P. Spiessens. Fine-grained parallel genetic algorithms. In J. D. Schaffer, editor, *Proceedings of the Third International Conference on Genetic Algorithms*, page 428. Morgan Kaufmann, 1989.

[134] O. L. Mangasarian, R. Setiono, and W.-H Goldberg. Pattern recognition via linear programming: Theory and application to medical diagnosis. In T. F. Coleman and Y. Li, editors, *Large-Scale Numerical Optimization*, pages 22–31. SIAM, 1990.

[135] D. Mange and M. Tomassini (Eds). *Bio-Inspired Computing Machines: Towards Novel Computational Architectures.* Presses Polytechniques et Universitaires Romandes, Lausanne, 1998.

[136] H. M. Markowitz. *Portfolio Selection.* John Wiley & Sons, New York, 1959.

[137] R. Maydole. Many-valued logic as a basis for set theory. *Journal of Philosophical Logic*, 4:269–291, 1975.

[138] H. A. Mayer. Symbiotic coevolution of artificial neural networks and training data sets. In A. Eiben, D. Bäck, M. Schoenauer, and H.-P. Schwefel, editors, *Parallel Problem Solving from Nature - PPSN V*, volume 1498 of *Lecture Notes in Computer Science*, pages 511–520. Springer-Verlag, Heidelberg, 1998.

[139] J. L. McCulloch and W. Pitts. A logical calculus of ideas immanent in nervous activity. *Bulletin of Mathematical Biophysics*, 5:115–133, 1943.

[140] C. Mead. *Analog VLSI and Neural Systems.* Addison-Wesley, May 1989.

[141] K. Menger. Statistical metric spaces. *Proceedings of the National Academy of Science of the USA*, 28:535–537, 1942.

[142] C. J. Merz and P. M. Murphy. The UCI repository of machine learning databases. URL: `http://www.ics.uci.edu/~mlearn/MLRepository.html`, 1996.

[143] Z. Michalewicz. *Genetic Algorithms + Data Structures = Evolution Programs, 3rd Edition.* Springer-Verlag, Berlin, 1996.

[144] O. Miglino, H. H. Lund, and S. Nolfi. Evolving Mobile Robots in Simulated and Real Environments. *Artificial Life*, 2:417–434, 1996.

[145] G. F. Miller, P. M. Todd, and S. U. Hegde. Designing neural networks using genetic algorithms. In J. D. Schaffer, editor, *Proceedings of the Third International Conference on Genetic Algorithms*, pages 379–384. Morgan Kaufmann, 1989.

[146] M. L. Minsky. *Computation: Finite and Infinite Machines.* Prentice-Hall, Englewood Cliffs, New Jersey, 1967.

[147] M. Misra. Parallel environments for implementing neural networks. *Neural Computing Surveys*, 1:48–60, 1997.

[148] T. M. Mitchell. *Machine Learning.* McGrow-Hill, New York, 1997.

[149] D. Montana and L. Davis. Training feedforward neural networks using genetic algorithms. In *Proceedings of the 11th International Conference on Artificial Intelligence*, pages 762–767. Morgan Kaufmann, 1989.

[150] N. Morgan, J. Beck, J. Kohn, P. Bilmes, J. Allman, and J. Beer. The Ring Array Processor (RAP): a multiprocessing peripheral for connectionist applications. *J. Parallel Distrib. Comput.*, 14:248–259, 1992.

[151] H. Mühlenbein and D. Schlierkamp-Voosen. The science of breeding and its application to the breeder genetic algorithm (BGA). *Evolutionary Computation*, 1(4):335–360, Winter 1993.

[152] H. Mühlenbein, M. Schomish, and J. Born. The parallel genetic algorithm as a function optimizer. *Parallel Computing*, 17:619–632, 1991.

[153] M. Murakawa, S. Yoshizawa, I. Kajitani, X. Yao, N. Kajihara, M. Iwata, and T. Higuchi. The GRD chip: Genetic reconfiguration of DSPs for neural network processing. *IEEE Transactions on Computers*, 48(6):628–639, June 1999.

[154] P. Nangsue and S. E. Conry. An agent-oriented, massively distributed parallelization model of evolutionary algorithms. In J. Koza, editor, *Late Breaking Papers, Genetic Programming 1998*, pages 160–168. Stanford University, 1998.

[155] D. Nauck. Neuro-fuzzy systems: Review and prospects. In *Fifth European Congress on Intelligent Techniques and Soft Computing (EUFIT'97)*, pages 1044–1053. 1997.

[156] D. Nauck, F. Klawonn, and R. Kruse. *Foundations of Neuro-Fuzzy Systems.* John Wiley & Sons, Chichester, 1997.

[157] D. Nauck and R. Kruse. A neuro-fuzzy approach to obtain interpretable fuzzy systems for function approximation. In *Proc. IEEE International Conference on Fuzzy Systems 1998 (FUZZ-IEEE'98)*, pages 1106–1111. IEEE Press, 1998.

[158] C. V. Negoita and D. A. Ralescu. *Applications of Fuzzy Sets to Systems Analysis.* Birkhauser, Basel, 1975.

[159] S. Nolfi and D. Floreano. Learning and evolution. *Autonomous Robots*, 7(1):89–113, 1999.

[160] E. Oja. Principal component analysis. In Michael A. Arbib, editor, *Handbook of Brain Theory and Neural Networks*, pages 753–756. MIT Press, 1995.

[161] M. Oussaidène, B. Chopard, O. Pictet, and M. Tomassini. Parallel genetic programming and its application to trading model induction. *Parallel Computing*, 23:1183–1198, 1997.

[162] Y.-H. Pao. Process monitoring amd optimization for power systems applications. In *Proceedings of the 1994 IEEE International Conference on Neural Networks*, page Part 6 (of 7). Orlando, FL, 1994.

[163] C. H. Papadimitriou and K. Steiglitz. *Combinatorial Optimization, Algorithms and Complexity.* Prentice-Hall, Englewood Cliffs, 1982.

[164] J. Paredis. Coevolutionary computation. *Artificial Life*, 2(4):355–375, 1995.

[165] K. L. Parker and A. L. Thornburg. Parallelized backpropagation training and its effectiveness. In *Proceedings of*

*the International Joint Conference on Neural Networks*, pages 179–182. Lawrence Erlbaum Associates, Hillsdale, 1990.

[166] H. Paugam-Moisy. Parallel neural computing based on network duplicating. In I. Pitas, editor, *Parallel Algorithms for Digital Image Processing, Computer Vision and Neural Networks*, pages 305–340. Wiley, New York, 1993.

[167] C. A. Peña-Reyes and M. Sipper. Evolving fuzzy rules for breast cancer diagnosis. In *Proceedings of 1998 International Symposium on Nonlinear Theory and Applications (NOLTA'98)*, volume 2, pages 369–372. Presses Polytechniques et Universitaires Romandes, Lausanne, 1998.

[168] C. A. Peña-Reyes and M. Sipper. A fuzzy-genetic approach to breast cancer diagnosis. *Artificial Intelligence in Medicine*, 17(2):131–155, October 1999.

[169] H. Pedrycz, I. Hayashi, and N. Wakami. A learning method of fuzzy inference rules by descent method. In *Proceedings of the IEEE Iternational Conference on Fuzzy Systems 1992*, pages 203–210. S. Diego, CA, 1992.

[170] W. Pedrycz. *Fuzzy Control and Fuzzy Systems.* John Wiley & Sons, New York, 1993.

[171] W. Pedrycz and H. C. Card. Linguistic interpretation of self-organizing maps. In *Proceedings of the IEEE Iternational Conference on Fuzzy Systems 1992*, pages 371–378. S. Diego, CA, 1992.

[172] W. Pedrycz and J. Valente de Oliveira. Optimization of fuzzy models. *IEEE Transactions on Systems, Man and Cybernetics*, 26(4):627–636, August 1996.

[173] W. Pedrycz and A. F. Rocha. Fuzzy-set based models of neurons and knowledge-based networks. *IEEE Transactions on Fuzzy Systems*, 1(4):254–266, 1993.

[174] A. Perez. Artificial neural networks: Algorithms and hardware implementation. In D. Mange and M. Tomassini, editors, *Bio-Inspired Computing Machines: Towards Novel Computational Architectures*, pages 289–316. Presses Polytechniques et Universitaires Romandes, Lausanne, 1998.

[175] A. Pétrovski and H. Paugam-Moisy. Parallel neural computation based on algebraic partitioning. In I. Pitas, editor, *Parallel Algorithms for Digital Image Processing, Computer Vision and Neural Networks*, pages 259–304. Wiley, New York, 1993.

[176] D. Polani and T. Uthmann. Training Kohonen feature maps in different topologies: an analysis using genetic algorithms. In S. Forrest, editor, *Proceedings of the Fifth International Conference on Genetic Algorithms*, pages 326–333. Morgan Kaufmann Publishers, San Mateo, California, 1993.

[177] R. Poluzzi, G. G. Rizzotto, and A. Tettamanzi. An evolutionary algorithm for fuzzy controller synthesis and optimization based on SGS-Thomson's W.A.R.P. fuzzy processor. In L. A. Zadeh E. Sanchez, T. Shibata, editor, *Genetic algorithms and fuzzy logic systems: Soft computing perspectives*, pages 71–89. World Scientific, Singapore, 1997.

[178] M. Potter and K. De Jong. A cooperative coevolutionary approach to function optimization. In *Procs. of the Third Conference on Parallel Problem Solving from Nature, Y. Davidor and H.-P. Schwefel (Eds.)*, pages 249–257. Lecture Notes in Computer Science Vol. 866, Springer-Verlag, Heidelberg, 1994.

[179] M. Potter and K. De Jong. Evolving neural networks with collaborative species. In *Procs. of the 1995 Summer Computer Simulation Conference*, pages 340–345. The Society for Computer Simulation, Ottawa, Canada, 1995.

[180] T. J. Procyk and E. H. Mamdani. A linguistic self-organising process controller. *Automatica*, 15, 1978.

[181] P. Prusinkiewicz and A. Lindenmayer. *The Algorithmic Beauty of Plants*. Springer-Verlag, New York, 1990.

[182] W. Punch. How effective are multiple populations in genetic programming. In J. R. Koza, W. Banzhaf, K. Chellapilla, K. Deb, M. Dorigo, D. B. Fogel, M. Garzon, D. Goldberg, H. Iba, and R. L. Riolo, editors, *Genetic Programming 1998: Proceedings of the Third Annual Conference*, pages 308–313. Morgan Kaufmann, San Francisco, CA, 1998.

[183] J. R. Quinlan. Induction on decision trees. *Machine Learning*, 1:81–106, 1986.

[184] H. Raza, P. Ioannou, and H. Youssef. Surface failure detection for an F/A-18 aircraft using neural networks and fuzzy logic. In *Proceedings of the 1994 IEEE International Conference on Neural Networks*, pages Part 5 (of 7), 3363–3368. Orlando, FL, 1994.

[185] I. Rechenberg. *Evolutionsstrategie: Optimierung technischer Systeme nach Prinzipien der biologischen Evolution.* Fromman-Holzboog Verlag, Stuttgart, 1973.

[186] C. R. Reeves and S. J. Taylor. Selection of training data for neural networks by a genetic algorithm. In A. Eiben, D. Bäck, M. Schoenauer, and H.-P. Schwefel, editors, *Parallel Problem Solving from Nature - PPSN V*, volume 1498 of *Lecture Notes in Computer Science*, pages 633–642. Springer-Verlag, Heidelberg, 1998.

[187] N. Rescher. *Many-valued Logic.* McGraw-Hill, New York, 1969.

[188] R. Rojas. *Theorie der Neuronalen Netze.* Springer-Verlag, Heidelberg, 1993.

[189] J. Rosca and D. Ballard. Discovery of subroutines in genetic programming. In P.J. Angeline and K.E. Kinnear Jr. (Eds.), editors, *Advances in Genetic Programming 2*, pages 177–201. Cambridge, Massachusetts, 1996.

[190] F. Rosenblatt. *Principles of Neurodynamics: Perceptrons and the theory of brain mechanics.* Spartan Books, Washington D.C., 1962.

[191] T.J. Ross. *Fuzzy Logic with Engineering Applications.* McGraw-Hill, New York, 1995.

[192] G. Rudolph. Parallel approaches to stochastic global optimization. In W. Joosen and E. Milgrom, editors, *Parallel Computing: from Theory to sound Practice. Proceedings of the European Workshop on Parallel Computing*, pages 256–267. IOS Press, Amsterdam, 1992.

[193] G. Rudolph. On correlated mutations in evolution strategies. In R. Männer and B. Manderick, editors, *Parallel Problem Solving from Nature 2*, pages 105–114. Elsevier, Amsterdam, 1992.

[194] G. Rudolph and J. Sprave. A cellular genetic algorithm with self-adjusting acceptance thereshold. In *In First IEE/IEEE International Conference on Genetic Algorithms in Engineering Systems: Innovations and Applications*, pages 365–372. IEE, London, 1995.

[195] A. Rueda and W. Pedrycz. Hierarchical fuzzy-neural PID controller for robot manipulators. In *Proceedings of the IEEE International Conference on Fuzzy Systems*, pages 673–676. IEEE Press, 1994.

[196] G. Salton. *Automatic Text Processing: The transformation, analysis, and retrieval of information by computer.* Addison-Wesley, Reading, MA, 1989.

[197] A. L. Samuel. Some studies in machine learning using the game of checkers. In E. A. Feigenbaum and J. Feldman, editors, *Computers and Thought.* McGraw-Hill, New York, 1959.

[198] E. Sanchez. Resolution of composite fuzzy relation equations. *Information and Control*, 30:38–48, 1976.

[199] E. Sanchez. Fuzzy genetic algorithms in soft computing environment. In *Fifth IFSA World Congress '93*, pages XLIV–L. Seoul, Korea, 1993.

[200] E. Sanchez and P. Pierre. Fuzzy logic and genetic algorithms in information retrieval. In *3rd International Conference on Fuzzy Logic, Neural Networks and Soft Computing.* 1994.

[201] J. Sarma and K. De Jong. An analysis of the effect of the neighborhood size and shape on local selection algorithms. In H.-M. Voigt, W. Ebeling, I. Rechenberg, and H.-P. Schwefel, editors, *Parallel Problem Solving from Nature - PPSN IV*, volume 1141 of *Lecture Notes in Computer Science*, pages 236–244. Springer-Verlag, Heidelberg, 1996.

[202] H.-P. Schwefel. Kybernetische Evolution als Strategie der experimentellen Forschung in der Strömungstechnik. Master's thesis, Technische Universität Berlin, Berlin, 1965.

[203] H.-P. Schwefel. *Numerical optimization of computer models.* Wiley, Chichester, New York, 1981.

[204] H.-P. Schwefel. Collective phenomena in evolutionary systems. In *Preprints of the 31st Annual Meeting of the International Society for General System Research*. Budapest, 1987.

[205] R. Setiono. Extracting rules from pruned neural networks for breast cancer diagnosis. *Artificial Intelligence in Medicine*, 8:37–51, 1996.

[206] R. Setiono and H. Liu. Symbolic representation of neural networks. *IEEE Computer*, 29(3):71–77, March 1996.

[207] G. Shafer. *A Mathematical Theory of Evidence*. Princeton University Press, Princeton, NJ, 1976.

[208] S. Shao. Fuzzy self-organizing controller and its application for dynamic processes. *Fuzzy Sets and Systems*, 26:151–164, 1988.

[209] A. A. Siddigi and S. M. Lucas. A comparison of matrix rewriting versus direct encoding for evolving neural networks. In *Proceedings of 1998 IEEE International Conference on Evolutionary Computation (ICEC'98)*, pages 392–397. IEEE Press, 1998.

[210] M. Sipper. *Evolution of Parallel Cellular Machines: The Cellular Programming Approach*. Springer-Verlag, Heidelberg, 1997.

[211] T. Starkweather, D. Whitley, and K. Mathias. Optimization using distributed genetic algorithms. In H.-P. Schwefel and R. Männer, editors, *Parallel Problem Solving from Nature*, volume 496 of *Lecture Notes in Computer Science*, pages 176–185. Springer-Verlag, Heidelberg, 1991.

[212] C. D. Stephens and N. Mackin. The validation of an orthodontic expert system rulebase for fixed appliance treatment planning. *European Journal of Orthodontics*, 20(5):569–578, October 1998.

[213] G. Stylios and O. J. Sotomi. Neuro-fuzzy control system for intelligent sewing machines. In *IEE Conference Publications, No. 395*, pages 241–246. IEE, Stevenage, UK, 1994.

[214] N. Sundararajan and P. Saratchandran. *Parallel Architectures for Artificial Neural Networks*. IEEE Press, Piscataway, NJ, 1998.

[215] I. Taha and J. Ghosh. Evaluation and ordering of rules extracted from feedforward networks. In *Proceedings of the IEEE International Conference on Neural Networks*, pages 221–226. 1997.

[216] T. Takagi and M. Sugeno. Fuzzy identification of systems and its applications to modeling and control. *IEEE Transactions on Systems, Man and Cybernetics*, 15(1):116–132, 1985.

[217] A. R. Tanguay and B. K. Jenkins. Optical architectures for neural network implementations. In Michael A. Arbib, editor, *Handbook of Brain Theory and Neural Networks*, pages 673–677. MIT Press, 1995.

[218] A. R. Tanguay and B. K. Jenkins. Optical components for neural network implementations. In Michael A. Arbib, editor, *Handbook of Brain Theory and Neural Networks*, pages 677–682. MIT Press, 1995.

[219] A. Tettamanzi. An evolutionary algorithm for fuzzy controller synthesis and optimization. In *IEEE International Conference on Systems, Man and Cybernetics*, volume 5/5, pages 4021–4026. IEEE Systems, Man and Cybernetics Society, 1995.

[220] A. Tettamanzi. Evolutionary algorithms and fuzzy logic: A two-way integration. In P. W. Wang, editor, *Proceedings of the 2nd Annual Joint Conference on Information Sciences*, pages 464–467. Duke University, Box 90291, Durham, NC, September 28–October 1, 1995.

[221] P. Thrift. Fuzzy logic synthesis with genetic algorithms. In R. K. Belew and L. B. Booker, editors, *Proceedings of the Fourth International Conference on Genetic Algorithms*, pages 509–513. Morgan Kaufmann, San Mateo, CA, 1991.

[222] M. Tomassini. The parallel genetic cellular automata: Application to global function optimization. In R. F. Albrecht, C. R. Reeves, and N. C. Steele, editors, *Proceedings of the International Conference on Artificial Neural Networks and Genetic Algorithms*, pages 385–391. Springer-Verlag, Heidelberg, 1993.

[223] L. H. Tsoukalas and R. E. Uhrig. *Fuzzy and Neural Approaches in Engineering.* John Wiley & Sons, New York, 1997.

[224] R. E. Uhrig, L. H. Tsoukalas, and A. Ikonomopoulos. Application of neural networks and fuzzy systems to power plants. In *Proceedings of the 1994 IEEE International Conference on Neural Networks*, pages Part 6 (of 7), 510–512. Orlando, FL, 1994.

[225] J. Urzelai and D. Floreano. Evolutionary robotics: coping with environmental change. In *Proceedings of the Genetic and Evolutionary Computation Conference GECCO 2000*, pages 941–948. Morgan Kaufmann, San Francisco, 2000.

[226] J. Urzelai and D. Floreano. Evolutionary robots with fast adaptive behavior in new environments. In T. C. Fogarty, J. Miller, A. Thompson, and P. Thomson, editors, *Third International Conference on Evolvable Systems: From Biology to Hardware (ICES2000)*, pages 241–251. Springer-Verlag, Berlin, 2000.

[227] N. Vadiee. Fuzzy rule based expert systems–I and II. In M. Jamshidi, N. Vadiee, and T. Ross, editors, *Fuzzy Logic and Control: Software and hardware applications*, Chapter 4. Prentice-Hall, Englewood Cliffs, N.J., 1993.

[228] M. Valenzuela-Rendón. The fuzzy classifier system: A classifier system for continuously varying variables. In R. K. Belew and L. B. Booker, editors, *Proceedings of the Fourth International Conference on Genetic Algorithms*, pages 346–353. Morgan Kaufmann, San Mateo, CA, 1991.

[229] E. Vittoz. Analog VLSI signal processing: Why, where and how? *Journal of VLSI Signal Processing*, 8:27–44, July 1994.

[230] H.-M. Voigt. Fuzzy evolutionary algorithms. Technical Report TR-92-038, International Computer Science Institute (ICSI), Berkeley, CA, 1992.

[231] H.-M. Voigt. Soft genetic operators in evolutionary algorithms. In W. Banzhaf and F. H. Eeckman, editors, *Evolution and Biocomputation*, pages 123–141. Springer-Verlag, Berlin, 1995.

[232] H.-M. Voigt and T. Anheyer. Modal mutations in evolutionary algorithms. In Z. Michalewicz, J. D. Shaffer, H.-P. Schwefel, D. B. Fogel, and H. Kitano, editors, *Proceedings of the First IEEE International Conference on Evolutionary Computation*, pages 88–92. IEEE Press, Piscataway, NJ, 1994.

[233] H.-M. Voigt, J. Born, and I. Santibañez Koref. A multivalued evolutionary algorithm. Technical Report TR-93-022, International Computer Science Institute (ICSI), Berkeley, CA, 1993.

[234] H.-M. Voigt, H. Mühlenbein, and D. Cvetković. Fuzzy recombination for the breeder genetic algorithm. In L. Eshelman, editor, *Proceedings of the Sixth International Conference on Genetic Algorithms*, pages 104–111. Morgan Kaufmann, San Francisco, CA, 1995.

[235] C. von der Malsburg. Self-organization of orientation sensitive cells in the striate cortex. *Kybernetik*, 14:85–100, 1973.

[236] N. Wakami, S. Araki, and H. Nomura. Recent applications of fuzzy logic to home appliances. In *Proceedings of the Industrial Electronics Conference (IECON)*, pages 155–160. IEEE Computer Society Press, Los Alamitos, CA, 1993.

[237] P. Wang. Approaching degree method. In *Fuzzy Set Theory and its Applications*. Science and Technology Press, Shanghai, 1983. In Chinese, cited in [191].

[238] A. S. Weigend and N. A. Gershenfeld (Eds.). *Times Series Prediction: Forecasting the Future and Understanding the Past.* Addison-Wesley, Reading, MA, 1994.

[239] P.J. Werbos. *The Roots of Backpropagation: From ordered derivatives to Neural Neworks and Political Forecasting.* John Wiley and Sons, New York, 1994.

[240] D. Whitley. Cellular genctic algorithms. In S. Forrest, editor, *Proceedings of the Fifth International Conference on Genetic Algorithms*, page 658. Morgan Kaufmann Publishers, San Mateo, California, 1993.

[241] D. Whitley. Genetic algorithms and neural networks. In M. Galan G. Winter, J. Périaux and P. Cuesta, editors, *Genetic Algorithms in Engineering and Computer Science*, pages 203–216. John Wiley, 1995.

[242] D. Whitley and T. Hanson. Optimizing neural networks using faster, more accurate genetic search. In J. D. Schaffer, editor, *Proceedings of the Third International Conference on Genetic Algorithms*, pages 391–396. Morgan Kaufmann, 1989.

[243] D. Whitley, S. Rana, and R. B. Heckendorn. Island model genetic algorithms and linearly separable problems. In D. Corne and J. L. Shapiro, editors, *Evolutionary Computing: Proceedings of the AISB Workshop, Lecture notes in computer science, vol. 1305*, pages 109–125. Springer-Verlag, Berlin, 1997.

[244] D. Whitley, T. Starkweather, and C. Bogart. Genetic algorithms and neural networks: optimizing connections and connectivity. *Parallel Computing*, 14:347–361, 1990.

[245] B. Widrow and M. Lehr. Perceptrons, Adalines, and Backpropagation. In Michael A. Arbib, editor, *Handbook of Brain Theory and Neural Networks*, pages 719–724. MIT Press, 1995.

[246] S. W. Wilson. Classifier systems based on accuracy. *Evolutionary Computation*, 3(2):149–175, 1995.

[247] H. Y. Xu and G. Vukovich. A fuzzy genetic algorithm with effective search and optimization. In *Proceedings of the 1993 International Joint Conference on Neural Networks*, pages 2967–2970. Nagoya, Japan, October 25–29, 1993.

[248] H. Y. Xu, G. Vukovich, Y. Ichikawa, and Y. Ishii. Fuzzy evolutionary algorithms and automatic robot trajectory generation. In Z. Michalewicz, J. D. Shaffer, H.-P. Schwefel, D. B. Fogel, and H. Kitano, editors, *Proceedings of the First IEEE International Conference on Evolutionary Computation*, pages 595–600. IEEE Press, Piscataway, NJ, 1994.

[249] R. R. Yager and D. P. Filev. *Essentials of Fuzzy Modeling and Control.* John Wiley & Sons, Inc., 1994.

[250] K. Yamazaki, H. Kaneko, S. Yamaguchi, K. Y. Watanabe, Y. Taniguchi, and O. Motojima. Design of the central control system for the large helical device (LHD). *Nuclear Instruments and Methods in Physics Research, Section A: Accelerators, Spectrometers, Detectors and Associated Equipment*, 352(1–2):43–46, 1994.

[251] X. Yao. Evolving artificial neural networks. *Proceedings of the IEEE*, 87(9):1423–1447, 1999.

[252] X. Yao and T. Higuchi. Promises and challenges of evolvable hardware. *IEEE Transactions on Systems, Man, and Cybernetics, Part C*, 29(1):87–97, 1999.

[253] X. Yao and Y. Liu. A new evolutionary system for evolving artificial neural networks. *IEEE Transactions on Neural Networks*, 8:694–713, 1997.

[254] X. Yao and Y. Liu. Making use of population information in evolutionary artificial neural networks. *IEEE Transactions on Systems, Man, and Cybernetics, Part B*, 28(3):417–425, 1998.

[255] H. Yoon and J.H. Nang. Multilayer neural networks on distributed memory multiprocessors. In *Proceedings of the International Neural Network Conference*, pages 669–672. Kluwer, Dordrecht, 1990.

[256] L. Zadeh. A theory of approximate reasoning. In J. Hayes, D. Michie, and L. Mikulich, editors, *Machine Intelligence*. Halstead Press, New York, 1979.

[257] L. A. Zadeh. Fuzzy sets. *Information and Control*, 8:338–353, 1965.

[258] L. A. Zadeh. Outline of a new approach to the analysis of complex systems and decision processes. *Information Science*, 9:43–80, 1973.

[259] L. A. Zadeh. The concept of a linguistic variable and its application to approximate reasoning, I–II. *Information Science*, 8:199–249, 301–357, 1975.

[260] L. A. Zadeh. Fuzzy sets as a basis for a theory of possibility. *Fuzzy Sets and Systems*, 1:3–28, 1978.

[261] L. A. Zadeh. The calculus of fuzzy if-then rules. *AI Expert*, 7(3):22–27, March 1992.

[262] L. A. Zadeh. Probability theory and fuzzy logic are complementary rather than competitive. *Technometrics*, 37:271–276, 1995.

[263] B. Zhang and H. Mühlenbein. Balancing accuracy and parsimony in genetic programming. *Evolutionary Computation*, 3:17–38, 1995.

[264] H.-J. Zimmermann. *Fuzzy Set Theory and its Applications*. Kluwer, Dordrecht, 2nd edition, 1991.

# Index

Production: Druckhaus Beltz, Hemsbach